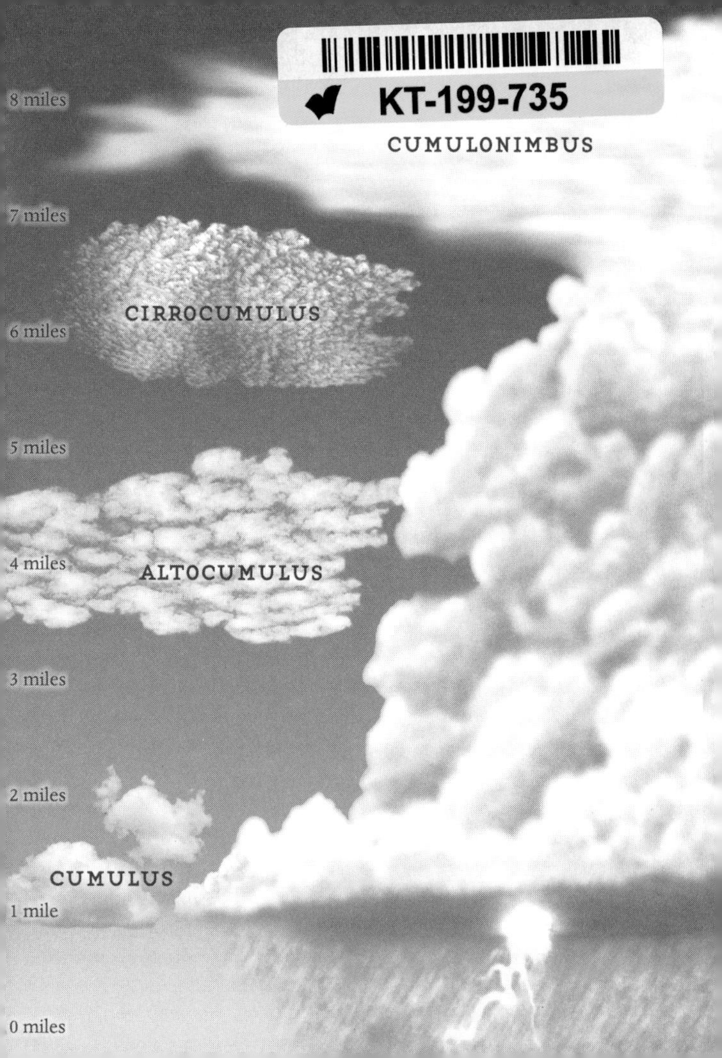

8 miles

CUMULONIMBUS

7 miles

CIRROCUMULUS

6 miles

5 miles

ALTOCUMULUS

4 miles

3 miles

2 miles

CUMULUS

1 mile

0 miles

THE CLOUD COLLECTOR'S HANDBOOK

This cloud collection
belongs to:

....

THE
CLOUD COLLECTOR'S
HANDBOOK

by Gavin Pretor-Pinney

An official publication of
THE CLOUD APPRECIATION SOCIETY
www.cloudappreciationsociety.org

∫

SCEPTRE

Image research by society Photo Editor, Ian Loxley (Member 1868).
Meteorological guidance by Stephen Burt (Member 2814)
of The Royal Meteorological Society.

HOW TO COLLECT CLOUDS

YOU MIGHT WELL THINK that cloud collecting sounds like a ridiculous idea. How can anyone collect such ephemeral and free-spirited things as clouds? Surely, they're just about as uncollectable as anything gets.

Magicked into being by the inscrutable laws of the atmosphere, clouds exist in a state of constant flux, shifting effortlessly from one form to another. One moment, they're joining and spreading into undulating layers. The next, they're breaking into torn shreds. One moment, they're building upwards in enormous, weighty towers with dark, brooding bases. The next, they're cascading back down in delicate, translucent streaks. And then they're gone – shedding their moisture as rain or just evaporating into the blue. They're like expressions on the face of the sky, and certainly not candidates for a display case. Given all the possible things you might consider collecting, clouds would seem to be a completely rubbish option.

But that's where you're wrong. You don't have to own something to collect it. You don't even have to hold it. You just have to notice it and record it.

And that is what this handbook is for. The entries will help you identify a whole range of distinctive cloud types, and some of the amazing optical effects produced by clouds as they scatter the sunlight. When you spot a particular cloud type, add it to your collection by noting down the details on the relevant page. Ideally, keep a camera to hand so that you can back up your claims with photographic evidence.

With each addition to your collection, you earn cloud-collecting points, which are determined by how hard each cloud or effect is to see. While a common old Stratocumulus

Altocumulus lenticularis, ruddy from another long day of being a beautiful cloud.

cloud only earns you 10 points, the fleeting crescent of a horseshoe vortex earns 50 points. The maximum score of 55 belongs to the rare and dramatic breaking waves of the Kelvin-Helmholtz cloud – the jewel of any cloud collection.

Your points should be entered religiously on the Contents & Scorecard on p7 and, as they mount up, they should be counted and re-counted with a greedy cackle. They'll be essential in judging the worth of your collection, and fuelling the bitter rivalry that will develop with fellow cloudspotters.

This handbook is intended to work as a complement to your own photographic records. Of course, you don't have to take pictures, but few cloudspotters can resist. All the photographs here were taken by members of The Cloud Appreciation Society, and can be seen much larger on the society's website, along with tips on cloud photography (see web address below). As the physical manifestation of your cloud collection, such photographic records of meteorological moments serve as something to rifle through and caress back at home.

The system for naming clouds is rather like that for plants and animals, and uses Latin terms to divide them up into different *genera*, *species* and *varieties*. Only the more distinctive and recognisable cloud types are included here. An overview of all the officially recognised classifications appears on pp104–105. Technical terms used in the cloud explanations are written in italics, and explained on pp100–103. Remember, cloudscapes usually contain a whole range of different cloud types, so don't expect them always to have the orderly, distinct forms of these images.

While it may not have the permanence of a collection of coins, nor the swapability of one of rare stamps, there's something honest about a collection of clouds. They embody the impermanence of the world around us. 'Nature', wrote Ralph Waldo Emerson, 'is a mutable cloud which is always and never the same.'

Gavin Pretor-Pinney,
The Cloud Appreciation Society
www.cloudappreciationsociety.org/collecting

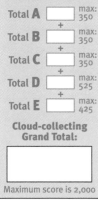

How Much is Your Collection Worth?

As you add clouds to your collection, update the totals in pencil from the scorecard opposite.

Total **A** ⬚ max: 350
+
Total **B** ⬚ max: 350
+
Total **C** ⬚ max: 350
+
Total **D** ⬚ max: 525
+
Total **E** ⬚ max: 425

Cloud-collecting Grand Total:

⬚

Maximum score is 2,000

CONTENTS & SCORECARD

Ten Main Cloud Types:

	Points	+ Bonus
Cumulus 9		
Stratocumulus 11		
Stratus 13		
Altocumulus 15		
Altostratus 17		
Cirrus 19		
Cirrocumulus 21		
Cirrostratus 23		
Nimbostratus 25		
Cumulonimbus 27		

Collecting Total **A**

Species & Varieties:

	Points	+ Bonus
Lenticularis 29		
Fibratus 31		
Castellanus 33		
Undulatus 35		
Lacunosus 37		
Radiatus 39		
Duplicatus 41		

Collecting Total **B**

Accessory Clouds & Supplementary Features:

	Points	+ Bonus
Pileus 43		
Velum 45		
Pannus 47		
Mamma 49		
Virga 51		
Arcus 53		
Tuba 55		
Incus 57		

Collecting Total **C**

Other Clouds:

	Points	+ Bonus
Cap & banner 59		
Kelvin-Helmholtz 61		
Contrails 63		
Pyrocumulus 65		
Roll cloud 67		
Fallstreak hole 69		
Horseshoe vortex 71		
Fog & mist 73		
Diamond dust 75		
Nacreous 77		
Noctilucent 79		

Collecting Total **D**

Cloud Optical Effects:

	Points	+ Bonus
Iridescence 81		
Corona 83		
Crepuscular rays 85		
Glory 87		
Rainbow 89		
Cloudbow 91		
22° halo 93		
Sundogs 95		
Circumzenithal arc 97		
Sun pillars 99		

Collecting Total **E**

Technical terms 100	
Cloud classification 104	
Photographers index ..106	
Cloud image index 108	
Index 110	
Credits 112	

Enter your totals in pencil, and copy them opposite for a running tally of your score.

Known as fair-weather clouds, Cumulus tend to appear on sunny days.

☐ **I spotted Cumulus**   Points scored: 15 + 15

Add to page 7

Date: .... Time: ....

Location: ....

Weather conditions: ....

☐ Photographed   Image file(s): ....

Cloud-collecting Points:
Any Cumulus .... 15 points ☐
Bonus for collecting all four species
(see opposite): humilis, mediocris,
congestus, fractus .... 15 points ☐

Typical altitudes: 1,000–5,000ft.
Precipitation: none, except for brief
showers from Cumulus congestus.
Don't confuse with: Stratocumulus (p11),
Altocumulus (p15), Cumulonimbus (p27).

8

Over The Isle of Hoy, Scotland, UK, by Eunice Clarke (Member 14190)

IF you've never spotted a Cumulus cloud, then you ought to get out more. This has to be one of the easiest types to add to your cloud collection (which explains why it earns a low score). Cumulus clouds are the cotton-wool puffs, with flat bases, that drift lazily across the sky on a sunny day. Generally forming a few hours after daybreak, they tend to dissipate before sundown, for they form on thermals – invisible columns of air rising from the ground as it is warmed by the Sun.

Most forms of Cumulus produce no rain or snow, and so are known as fair-weather clouds. But in *unstable air*, their bright, crisp cauliflower mounds can build upwards so that they develop from the small humilis *species*, through mediocris to the largest form, Cumulus congestus. With its ominous, shadowy base, this cloud is no longer fair-weather. Congestus can produce brief but sizeable showers, and can keep growing into fierce Cumulonimbus storm clouds (p27).

The little ones, by contrast, are only scary when they take the form of David Hasselhoff.

Top: A Cumulus congestus tower. *Bottom:* Cumulus fractus (as it forms/evaporates).

Top: over Mississauga, Ontario, Canada, by Bob Hotte (Member 13770)
Bottom: over Alberta, Canada, by Janice Smith (Member 10496)

Cumulus Species:	☐ **Fractus:** broken, with ragged edges.
☐ **Humilis:** wider than it is tall.	**Cumulus Varieties:**
☐ **Mediocris:** as tall as it is wide.	☐ **Radiatus:** lined up in 'cloud streets'
☐ **Congestus:** taller than it is wide.	(p39).

The ever-varied and omnipresent tones of Stratocumulus clouds.

| ☐ **I spotted Stratocumulus** | Points scored: 10 + 10 | Add to page 7 |

Date: .... Time: ....

Location: ....

Weather conditions: ....

☐ Photographed Image file(s): ....

Cloud-collecting Points:
Any Stratocumulus .... 10 points ☐
Bonus for when sunlight streams down
through holes in the layer, like huge
torch beams (see p85) ..... 10 points ☐

Typical altitudes: 1,000–4,500ft.
Precipitation: occasionally light rain,
snow or snow pellets.
Don't confuse with: Cumulus (p9),
Stratus (p13), Altocumulus (p15).

10

Over The Marmolada, Italy, by Claudia Harsch (Member 9596)

STRATOCUMULUS

THE most widespread of all cloud types, Stratocumulus is a low layer or patch of cloud that has a well-defined, clumpy base. The patches are either joined up, or have gaps in between. When the sky is overcast, and the cloud base appears to be low, with tones from white to dark grey, cloudspotters can confidently add Stratocumulus to their cloud collections.

Top: From above, Stratocumulus cities in the sky. *Bottom:* A high Stratocumulus.

High Stratocumulus that have *cloudlets* with gaps in between – a variety known as perlucidus – can be confused with the mid-level cloud Altocumulus (p15). But Stratocumulus is usually less orderly in appearance and its cloudlets are bigger (appearing larger than the width of three fingers, held at arm's length, when they are more than 30° above the horizon).

Due to its sun-blocking tendencies, Stratocumulus may not be the most popular cloud, but it is one of the most varied.

Stratocumulus Species: ☐ **Stratiformis:** extends over large areas of sky, rather than forming in patches. ☐ **Lenticularis:** smooth, lens-shaped mass (p29). ☐ **Castellanus:** top of the layer rises in turrets (p33). **Stratocumulus Varieties:** ☐ **Translucidus:** thin enough to show the outline of the Sun or Moon.	☐ **Perlucidus:** gaps between clumps. ☐ **Opacus:** thick enough to mask the Sun or Moon completely. ☐ **Duplicatus:** more than one layer, sometimes partly merged (p41). ☐ **Undulatus:** wave-like (p35). ☐ **Radiatus:** lined-up clumps that converge towards horizon (p39). ☐ **Lacunosus:** layer contains large holes, fringed with cloud (p37).

'I'd like to cancel my booking for the penthouse restaurant, please.'

☐ **I spotted Stratus**   Points scored: 15 + 20   Add to page 7

Date: .... Time: ....

Location: ....

Weather conditions: ....

☐ Photographed   Image file(s): ....

Cloud-collecting Points:
Any Stratus .... 15 points ☐
Bonus for when you are able to look
down on to Stratus and notice its
undulating upper surface.. 20 points ☐

Typical altitudes: 0–1,500ft.
Precipitation: just occasional drizzle, light snow or snow grains.
Don't confuse with: Altostratus (p17), Cirrostratus (p23), Nimbostratus (p25).

12

Over New York City, NY, US, by Zoltan Farago (Member 12528)

STRATUS

THE lowest-forming of all the cloud types, Stratus can give you a strangely claustrophobic feeling, even though you're outside. It is a featureless, grey overcast layer, which lurks around with its base generally no higher than 1,500ft from

Stratus fractus on a mountainside.

the ground. This is much lower than its equally charisma-free cousin, the Altostratus cloud (p17). Stratus can sometimes obscure the tops of tall buildings. When a cloud like this forms so low that it is at ground level, it is known as fog or mist. Since fog can sometimes form in a different way from airborne Stratus, it has a page of its own (p73).

One way that Stratus forms is when moist air cools as it blows over a relatively cold surface, such as a cold sea or land covered in thawing snow ('advection fog' (see p73) is formed in the same way when winds are gentler). Another is when air cools as it rises. This might be as it blows up the lower slopes of a mountainside or as warmer air slowly rides up over a region of colder (denser) air. Finally, Stratus can appear when fog, which has formed overnight, lifts from the ground as it is stirred by a freshening wind.

Over Bolungarvik, Iceland, by Michèle Gruber (Member 11072)

Stratus Species:	Stratus Varieties:
☐ **Nebulosus:** a grey, featureless layer – by far the most common form.	☐ **Opacus:** thick enough to mask the Sun or Moon completely.
☐ **Fractus:** patches or broken wisps, e.g. on hillsides. Known as pannus (p47) when forms in damp air below rain clouds.	☐ **Translucidus:** thin enough to show the outline of the Sun or Moon.
	☐ **Undulatus:** surface of the layer has a wave-like appearance. Since the cloud layer is so diffuse, this variety is very rarely observed (p35).

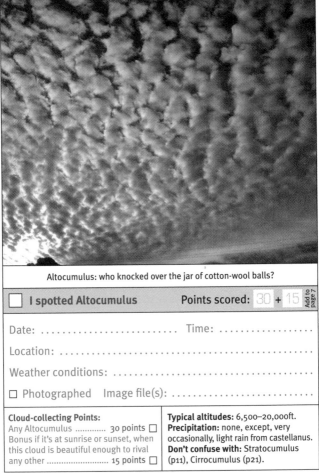

Altocumulus: who knocked over the jar of cotton-wool balls?

☐ **I spotted Altocumulus** Points scored: 30 + 15 Add to page 7

Date: .... Time: ....

Location: ....

Weather conditions: ....

☐ Photographed   Image file(s): ....

Cloud-collecting Points:	**Typical altitudes:** 6,500–20,000ft.
Any Altocumulus .... 30 points ☐	**Precipitation:** none, except, very occasionally, light rain from castellanus.
Bonus if it's at sunrise or sunset, when this cloud is beautiful enough to rival any other .... 15 points ☐	**Don't confuse with:** Stratocumulus (p11), Cirrocumulus (p21).

14

ALTOCUMULUS

THESE are typically mid-level layers or patches of *cloudlets,* which form clumps or rolls. They are white or grey, and shaded on the side away from the Sun. This distinguishes Altocumulus from the shade-free cloudlets of Cirrocumulus (p21). Another's the size of its cloudlets. These appear between the width of one and three fingers, held at arm's length, when they're more than 30° above the horizon.

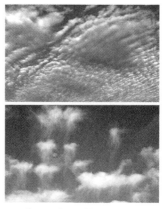

Top: Altocumulus undulatus. *Bottom:* Jellyfish trails, known as virga (p51).

The species of Altocumulus that stands out from all the others is lenticularis, described on p29. Rather than a layer of cloudlets, it has the form of large, smooth individual clouds.

With no less than four possible species and seven varieties, Altocumulus clouds produce the most dramatic and beautiful cloudscapes, especially in the rays of a low Sun.

Altocumulus Species:
☐ **Stratiformis:** extends over large areas of sky, rather than forming in patches.
☐ **Lenticularis:** smooth, lens-shaped mass (p29).
☐ **Castellanus:** top of the layer rises in turrets (p33).
☐ **Floccus:** Cumulus-like cloudlets, with ragged bases, often with virga (p51).
Altocumulus Varieties:
☐ **Translucidus:** thin enough to show the outline of the Sun or Moon.
☐ **Perlucidus:** gaps between clumps.
☐ **Opacus:** thick enough to mask the Sun or Moon completely.
☐ **Duplicatus:** more than one layer, sometimes partly merged (p41).
☐ **Undulatus:** wave-like (p35).
☐ **Radiatus:** lined-up clumps that converge towards horizon (p39).
☐ **Lacunosus:** layer contains large holes, fringed with cloud (p37).

15

Extra points for managing to stay awake when you watch Altostratus.

| ☐ **I spotted Altostratus** | **Points scored:** 15 + 10 | Add to page 7 |

Date: .... Time: ....

Location: ....

Weather conditions: ....

☐ Photographed Image file(s): ....

Cloud-collecting Points:	**Typical altitudes:** 6,500–16,500ft. The
Any Altostratus .... 15 points ☐	cloud base lowers as it thickens.
Bonus for managing to persuade	**Precipitation:** prolonged, but light.
anyone else to take the slightest bit of	**Don't confuse with:** Stratus (p13),
interest in this cloud .... 10 points ☐	Cirrostratus (p23).

16

ALTOSTRATUS

IT feels wrong to devote as much space to the rather drab and featureless Altostratus cloud as to its relative, the gloriously varied Altocumulus. Few cloudspotters will be seen to punch the air and high-five upon adding this one to their cloud collection. Altostratus is,

Streaks, known as undulatus radiatus.

after all, generally considered the most boring of all the cloud types. Although, even to say that, makes it sound rather more noteworthy than it deserves.

Altostratus is a mid-level, generally featureless, grey, overcast layer – a Tupperware sky that often extends over several thousand square miles. True to its dull nature, Altostratus produces little more than a lingering drizzle or light snow. Once it is thick enough to produce more significant precipitation, it has generally developed into the Nimbostratus cloud (p25).

The most common way for Altostratus to form is by the thickening of high Cirrostratus (p23), when a large region of warmer air pushes against one of colder air. The warmer air, being less dense, rises gently en masse over the colder.

Generally darker than Stratus (p13), Altostratus never produces *halo phenomena*, as Cirrostratus does. When you can see the Sun, it appears as if through ground glass.

There are no Altostratus Species. **Altostratus Varieties:** ☐ **Translucidus:** thin enough to show the outline of the Sun or Moon. ☐ **Opacus:** thick enough to mask the Sun or Moon completely.	☐ **Duplicatus:** more than one layer, visible only in the glancing light of a low Sun **(p41).** ☐ **Undulatus:** wave-like (p35). ☐ **Radiatus:** rows that converge towards horizon; very ocasional form (p39).

Cirrus clouds are streaks of falling ice crystals. This is the hooked species, uncinus.

| ☐ **I spotted Cirrus** | Points scored: 20 + 15 | Add to page 7 |

Date: .... Time: ....

Location: ....

Weather conditions: ....

☐ Photographed   Image file(s): ....

Cloud-collecting Points:
Any Cirrus .... 20 points ☐
Bonus for the classic uncinus
species with its distinctive hooked
formation (above) .... 15 points ☐

Typical altitudes: 20,000–40,000ft.
Precipitation: none reaching ground.
Don't confuse with: Altocumulus
(p15) that is producing trails of
precipitation, known as virga (p51).

18

Over Glen Coe, Highlands, Scotland, by Frank Howie (Member 4613)

CIRRUS

THE most ethereal looking of all the main types, Cirrus clouds are also the highest – composed entirely of ice crystals. These typically fall through the high winds of the upper *troposphere* to appear as delicate, celestial brush strokes, known as 'fallstreaks'. Cirrus often look like white locks of hair (from which the Latin name is derived).

Top: Twisted tousles of Cirrus intortus.
Bottom: Cirrus spissatus lingers after a Cumulonimbus (p27) has dissipated.

Cirrus clouds thickening and spreading across the blue can be the first signs of moisture developing at high altitudes, indicating the start of a common cloud progression that leads to Nimbostratus (p25) and produces rain or snow in a day or so.

Apart from the thick Cirrus spissatus (see below), all other forms of Cirrus can *refract* and reflect the sunlight to produce coloured arcs and rings, called *halo phenomena* (see pp93–99).

Cirrus Species: ☐ **Fibratus:** individual filaments without hooks or clumps at the end (p31). ☐ **Uncinus:** fallstreaks are shaped like hooks or commas. ☐ **Spissatus:** thick patches – sometimes as the anvil from an old Cumulonimbus (p27). ☐ **Castellanus:** fallstreaks from tiny tufts with turreted tops (p33).	☐ **Floccus:** fallstreaks from individual rounded tufts. **Cirrus Varieties:** ☐ **Intortus:** irregular, tangled fallstreaks. ☐ **Radiatus:** parallel filaments, usually aligned to the wind, converge towards horizon (p39). ☐ **Vertebratus:** filaments look like a fish skeleton, with a central spine. ☐ **Duplicatus:** filaments or streaks in more than one layer (p41).

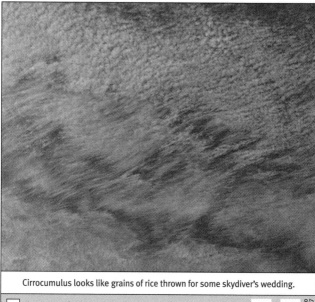

Cirrocumulus looks like grains of rice thrown for some skydiver's wedding.

☐ **I spotted Cirrocumulus**   Points scored: 40 + 10

Add to page 7

Date: ....   Time: ....

Location: ....

Weather conditions: ....

☐ Photographed   Image file(s): ....

Cloud-collecting Points:
Any Cirrocumulus .... 40 points ☐
Bonus for when it's near Cirrus (p19)
and patches of Cirrostratus (p23), since
they often form together .. 10 points ☐

Typical altitudes: 25,000–35,000ft.
Precipitation: none reaching ground.
Don't confuse with: the larger
cloudlets of the mid-level Altocumulus
(p15) – a much more common cloud.

20

Over Stratfield Mortimer, Berkshire, by Stephen Burt (Member 2814)

CIRROCUMULUS

THESE are high patches or layers of *cloudlets* that appear tiny, on account of their distance from the ground.

The best way to distinguish Cirrocumulus from lower Altocumulus (p15) is the size of the cloudlets, as well as the area of the sky covered by the layer as a whole. Being such a distance from the ground (often in the region of six miles), the cloudlets of Cirrocumulus appear so small that you often have to look

Cirrocumulus appears more often in patches than spread across the sky.

carefully to notice the cloud's grainy texture. For the cloud to be Cirrocumulus, these cloudlets must appear no larger than the width of a finger, held at arm's length, when they are more than 30° above the horizon.

Composed almost entirely of ice crystals, Cirrocumulus is actually a rare and fleeting cloud. Mostly, when you see a layer of cloudlets, they belong to the lower water-droplet cloud, Altocumulus. That's why Cirrostratus is a big points earner, scoring higher than any of the other main cloud types.

Cirrocumulus Species:
☐ **Stratiformis:** extends over large areas of sky, rather than forming in patches.
☐ **Lenticularis:** smooth, lens-shaped mass (p29).
☐ **Castellanus:** tops of cloudlets have turrets (p33) – you'll certainly need

binoculars to see this.
☐ **Floccus:** Cumulus-like cloudlets, with ragged bases, often with virga (p51).
Cirrocumulus Varieties:
☐ **Undulatus:** wave-like (p35).
☐ **Lacunosus:** layer contains large holes, fringed with cloud (p37).

Cirrostratus consists of ice crystals and, like this one, may appear fibrous.

☐ **I spotted Cirrostratus**   Points scored: 20 + 20   Add to page 7

Date: .... Time: ....

Location: ....

Weather conditions: ....

☐ Photographed   Image file(s): ....

Cloud-collecting Points:
Any Cirrostratus .... 20 points ☐
Bonus for noticing Cirrostratus producing more than one halo phenomenon
(see pp93–97) at a time ... 20 points ☐

Typical altitudes: 16,500–30,000ft.
Precipitation: none.
Don't confuse with: lower and thicker
Stratus (p13), Altostratus (p17), which
don't form halo phenomena.

CIRROSTRATUS

CIRROSTRATUS is a subtle, understated cloud that can easily go unnoticed – except, that is, by cloudspotters, keen to complete their collection of the ten main cloud types.

A delicate layer of ice crystals, often spread over vast areas of the sky, Cirrostratus can appear as no more than a light, milky whitening of the blue. It can sometimes look stripy or fibrous (the species known as fibratus) but more commonly lacks any variation in tone.

Top: A halo like this indicates Cirrostratus cloud. *Bottom:* A milky veil.

It also distinguishes itself as the best of the high clouds at producing *halo phenomena* (see pp93–99). A range of arcs, rings and points of light can appear as the sunlight is *refracted* and reflected by its tiny ice crystals. These don't always appear but, when they do, can exhibit beautiful rainbow colours. The presence of haloes is a sure way to distinguish Cirrostratus from Altostratus, which, being lower and consisting (at least partially) of droplets, doesn't produce them.

Cirrostratus Species:	Cirrostratus Varieties:
☐ **Fibratus:** made of delicate, parallel fibres (p31). ☐ **Nebulosus:** smooth, with no variation in tone.	☐ **Duplicatus:** more than one layer at different altitudes; very hard to distinguish unless differing winds at each altitude cause fibratus stripes to point in different directions (p41). ☐ **Undulatus:** wave-like (p35).

A lovely day by the seaside, courtesy of the Nimbostratus cloud.

| ☐ **I spotted Nimbostratus** | Points scored: 10 + 5 | Add to page 7 |

Date: .... Time: ....

Location: ....

Weather conditions: ....

☐ Photographed   Image file(s): ....

Cloud-collecting Points:
Any Nimbostratus .... 10 points ☐
Bonus (to make you feel better) if the
Nimbostratus turns your weekend at
the beach into a washout ... 5 points ☐

Typical altitudes: 0–10,000ft.
Precipitation: you bet.
Don't confuse with: Stratus (p13),
Altostratus (p17), Cumulonimbus (p27)
if you are directly below it.

24

Over Whitby, Yorkshire, UK, by Ian Loxley (Member 1868)

NIMBOSTRATUS

WHEN people claim clouds are depressing, they're often thinking of Nimbostratus. This thick, grey, featureless rain cloud gives all the other ones a bad name. Not only does it block much of the Sun's rays, casting everything in a dim, miserable light, it also produces rain – and lots of it.

These clouds can also produce snow.

Nimbostratus is one of only two cloud types that are defined as always producing rain or other precipitation. The other is the Cumulonimbus storm cloud (p27). From below, both appear as dark and ominous skies, but they can be distinguished by the nature of their precipitation. Compared with the brief heavy showers from individual Cumulonimbus clouds, the precipitation from Nimbostratus is much more steady, and can last for many hours.

Surreptitiously and without fanfare is how the Nimbostratus arrives. It generally results from the thickening and lowering of Altostratus (p17). Since one cloud leads to the other, the point of distinction between Alto- and Nimbostratus is rather academic. But when the cloud is dark, and the rain moderate to heavy, and its diffused base shows darker ragged patches of Stratus fractus, which is also known as pannus (p47), you can confidently add Nimbostratus to your cloud collection.

But you're unlikely to win any awards for your photos of this less than handsome cloud.

Over Glossop, Derbyshire, UK, by Dave Leech (Member 12529)

Nimbostratus Species & Varieties: Being such a featureless wet blanket of	a cloud, Nimbostratus is not considered to have any species or varieties.

The Cumulonimbus, King of Clouds, can sometimes grow to over ten miles high.

☐ **I spotted Cumulonimbus**   **Points scored:** 40 + 15 Add to page 7

Date: .... Time: ....

Location: ....

Weather conditions: ....

☐ Photographed   Image file(s): ....

Cloud-collecting Points:	**Typical altitudes:** bases around
Any Cumulonimbus .... 40 points ☐	2,000ft, can reach up to 45,000ft.
Bonus for when the King of Clouds	**Precipitation:** heavy showers, often hail.
is producing thunder and	**Don't confuse with:** Cumulus
lightning .... 15 points ☐	congestus (p9), Nimbostratus (p25).

26

CUMULONIMBUS

NO cloud collection is complete without the big one, the Godfather of clouds: Cumulonimbus. This enormous storm cloud, which is often in the shape of a blacksmith's anvil, can form individually or co-ordinate with neighbours to form *multicell* and *supercell storms*.

Cloudspotters should note that the anvil shape is visible only when looking at the cloud from many miles away. It develops from Cumulus congestus (p9), and is a

Top: A bad hair day for a Cumulonimbus capillatus. *Bottom:* View from below.

Cumulonimbus once its summit has changed from droplets to ice crystals – developing softer edges. Below a Cumulonimbus, you will see just its dark, ragged underside, which (being so low) appears to cover the whole sky. Distinguish it from Nimbostratus by how its moisture falls (see p25), and the fact that it produces thunder, lightning and often hail.

Cumulonimbus also gives rise to a whole range of *accessory clouds* and *supplementary features*, such as incus (p57), mamma (p49), pileus (p43), velum (p45), arcus (p53) and tuba (p55).

Cumulonimbus Species:	
(Only distinguishable when the cloud is in many miles away.) ☐ **Calvus:** top is of soft mounds, and isn't fibrous or striated (means 'bald' in Latin).	☐ **Capillatus:** developing from calvus into mature phase, top spreads out into familiar anvil plume of Cirrus-like fibres or striations (means 'hairy' in Latin). **There are no Cumulonimbus Varieties.**

Top: over Calcutt, Somerset, UK, by Ron Westmaas (Member 4451)
Bottom: over Carmignano, Italy, by Bernardo Herd-Smith (Member 14192)

Altocumulus lenticularis stacked in pile d'assiettes formation.

☐ **I spotted lenticularis**   Points scored: 45 + 25   Add to page 7

Date: . . . .   Time: . . . .

Location: . . . .

Weather conditions: . . . .

☐ Photographed   Image file(s): . . . .

Cloud-collecting Points:
Any lenticularis .... 45 points ☐
Bonus for the rare pile d'assiettes,
formed due to alternating layers of
moister and drier air .... 25 points ☐

Species is found in: Stratocumulus (p11), Altocumulus (p15), Cirrocumulus (p21).
Don't confuse with: pileus (p43), which only differs in that it forms over Cumulus clouds, not hills.

28

LENTICULARIS

LENTICULARIS clouds are contenders for the Weirdest-Looking-Clouds-in-the-Sky awards. Their name is Latin for a lentil, on account of their very distinctive lens shapes. They often look remarkably like flying saucers. Presumably, when they were named, no one could think of the Latin word for 'shaped like a UFO'.

Lenticularis species of high Cirrus cloud *(top)* and low Stratocumulus *(bottom)*.

Lenticularis can be found at low, medium and high *cloud levels*, although the most striking and dramatic ones tend to be the mid-level Altocumulus lenticularis. At whatever altitude they form, they are usually caused by a moist airstream flowing over raised ground, such as a hill or mountain peak. When the atmosphere in the area is *stable*, the air can develop a wave-like motion downstream, invisibly rising and dipping in the lee of the peak. If the air rises and cools enough, lenticularis clouds can appear at the crests of these waves. Unlike most clouds that drift along with the breeze, these hover even in the strongest winds (so long as air speed remains constant). Their positions in the airstream remain fixed, like the stationary waves of water behind a boulder in the current of a fast-moving stream.

When the airstream contains layers of moist air separated by drier air, a stacked formation can appear, known as 'pile d'assiettes' (which is French for 'your turn for the washing up').

Top: over Watlington, Oxfordshire, UK, by Richard B. Machin (Member 10911)
Bottom: over Omiš, Croatia, by Dunja Lučić (Member 14199)

Cirrus fibratus stretching off to the distance are also known as radiatus.

| ☐ **I spotted fibratus** | Points scored: 15 + 10 | Add to page 7 |

Date: . . . . Time: . . . .

Location: . . . .

Weather conditions: . . . .

☐ Photographed   Image file(s): . . . .

Cloud-collecting Points:
Any fibratus .... 15 points ☐
Bonus for when the hair-like strands
have been blown into a fashionable,
wavy style .... 10 points ☐

Species is found in: Cirrus (p19),
Cirrostratus (p23).
Don't confuse with: uncinus and
floccus species of Cirrus clouds (see
opposite and p19).

Over Norwich, East Anglia, UK, by Andrew Bluemel (Member 12532)

FIBRATUS

THE high, ice-crystal clouds of Cirrus (p19) and Cirrostratus (p23) are called fibratus when they have been drawn out by the wind into long, fine filaments. These close strands of cloud appear rather like hair run through with a comb. Such an orderly atmospheric hairstyle depends on high, continuous winds. These are more common up at Cirrus and Cirrostratus level, since the higher you climb through the *troposphere*, the faster the average wind speed becomes,

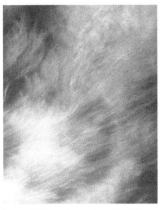

Father Christmas's beard, or the delicate strands of Cirrostratus fibratus.

and the less the wind is messed about by the influence of the ground.

The way to distinguish fibratus from the other Cirrus species that can also have somewhat parallel filaments, floccus and uncinus (see p19), is to look at the ends of the strands. In fibratus, the filaments do not descend from the fluffy tufts of cloud found in floccus, nor do they curve down from thicker heads to give the hooked, comma-like appearance of uncinus. Fibratus are simply thin, delicate strands of high cloud.

As expressions on the face of the sky, clouds can be indicators of the atmosphere's moods, but not so in the case of fibratus clouds. Other than indicating high, continuous winds up at cloud level, they tell nothing of the weather in store.

Perhaps they are just there to look nice.

Altocumulus clouds are the best ones to look at for the turreted tops of castellanus.

| ☐ **I spotted castellanus** | **Points scored:** 25 + 20 |

Add to page 7

Date: .... Time: ....

Location: ....

Weather conditions: ....

☐ Photographed   Image file(s): ....

Cloud-collecting Points:
Any castellanus .... 25 points ☐
Bonus for correctly predicting showers later in the day by spotting some at the Altocumulus level .... 20 points ☐

Species is found in: Stratocumulus (p11), Altocumulus (p15), Cirrus (p19), Cirrocumulus (p21).
Don't confuse with: floccus, whose cloudlets lack pronounced turrets.

Over Start Bay, Devon, UK, by Celia Warren (Member 10606)

CASTELLANUS

WHEN a layer of cloud rises in distinct turrets with bumpy tops that resemble crenellations, it is of the species known as castellanus – and this one can give an early indication of unsettled weather to come later in the day.

Aerial fortifications, called castellanus.

The turrets of castellanus can be found at all three *cloud levels*, but the ones that are hardest to identify are in the high clouds, Cirrocumulus (p21) and Cirrus (p19). As much as anything, this is because the cloud elements are so far away that they appear tiny from the ground, making any observation of the subtle nature of their tops rather challenging. Luckily, these examples of castellanus are also the least indicative of unsettled weather.

The low cloud, Stratocumulus (p11), may be described as castellanus when at least some of the cauliflower mounds that form its upper surface have grown taller than they are wide. These rising turrets can sometimes continue to grow upwards and develop into rain-bearing Cumulus congestus (p9), or even Cumulonimbus (p27) storm clouds.

But it is in the mid-level cloud, Altocumulus (p15), that the jagged crenellations of castellanus are most recognisable, and also where they best forecast unsettled weather. The vigorous turrets show that the air at the mid-cloud level is *unstable*. Any Cumulus that start to develop on thermals will, upon reaching this layer, continue growing with extra vigour, and quite possibly develop into tall, stormy Cumulonimbus clouds.

33

Like ripples in the sand: Cirrocumulus undulatus.

| ☐ **I spotted undulatus** | Points scored: 20 + 10 | Add to page 7 |

Date: . . . .   Time: . . . .

Location: . . . .

Weather conditions: . . . .

☐ Photographed   Image file(s): . . . .

Cloud-collecting Points:
Any undulatus . . . . 20 points ☐
Bonus for undulations at more than
one level, giving a crisscrossed effect
(see duplicatus, p41) . . . . 10 points ☐

Variety is found in: Stratus (p13), Stratocumulus (p11), Altocumulus (p15), Altostratus (p17), Cirrocumulus (p21), Cirrostratus (p23), noctilucent (p79).

34

UNDULATUS

WHEN the surface of a cloud layer, or the arrangement of its *cloudlets*, develops an undulating appearance that suggests waves, it's defined as the undulatus variety.

Top: Altocumulus undulatus. Bottom: Stratocumulus undulatus.

Waves and clouds have always had a close relationship. The interaction of currents in the atmosphere, and the effects of the terrain on the passage of winds, can result in a whole range of undulating currents of air. Generally, these are invisible, unless the rising parts of the undulations cool the air enough to produce clouds of droplets or ice crystals, which are thinner or absent in the sinking parts of the undulations. In such circumstances, the waves show up on the surface of the cloud or as cloud billows with gaps in between.

Undulatus usually forms when the air above and below the cloud layer is moving at differing speeds and / or in different directions. It is the shearing effect of the two airstreams that gives rise to the cloud billows, which resemble ripples on a sandy beach caused by the movement of water.

Wave formations in clouds are so common that the undulatus variety is found in six of the ten main cloud types (see opposite). Their presence is a reminder, to any who might forget, that the atmosphere around us is just as much an ocean as is the sea below.

Top: over Lastagneto Carducci, Iuscany, Italy, by Katrin Pfeifer (Member 12533)
Bottom: over The Barkly Tableland, Northern Territory, Australia, by Peter Mann (Member 14193)

A crater-pocked landscape, courtesy of Altocumulus lacunosus cloud holes.

| ☐ **I spotted lacunosus** | Points scored: 40 + 35 |

Add to page 7

Date: .... Time: ....

Location: ....

Weather conditions: ....

☐ Photographed   Image file(s): ....

Cloud-collecting Points:
Any lacunosus .... 40 points ☐
Bonus for when you spot the rarer
lacunosus among the unruly form of
Stratocumulus (p11) .... 35 points ☐

Variety is found in: Stratocumulus (p11), Altocumulus (p15), Cirrocumulus (p21), noctilucent (p79).
Don't confuse with: fallstreak holes (p69).

36

Over Worcester, Worcestershire, UK, by Kate Brookes (Member 9962)

LACUNOSUS

A rare, fleeting formation, the lacunosus variety is identified in terms of the gaps between cloud elements, rather than the clouds themselves. It is when a cloud layer is composed of more or less regular holes, around which fringes of cloud form, like a net or rough honeycomb.

Even though lacunosus forms at all three *cloud levels*, it is an elusive prize for any cloud collector, since it is so short-lived. Like the equally transient pileus cloud (p43), lacunosus therefore earns considerable cloud-collecting points.

Hole size varies from high Cirrocumulus (*top*) to low Stratocumulus (*bottom*).

The holes of this variety are formed by sinking pockets of air, and the cloud fringes around them by air rising up between the pockets to replace them. Such sinking can occur when a layer of cooler air finds itself over a warmer one. Being more dense, the cooler air sinks down through the warmer air. The appearance is similar to the rough honeycomb pattern you occasionally see on the surface of a hot cup of tea. As the tea on the surface cools and contracts, it sinks in pockets through the hotter tea below, which bubbles up in between to replace it. That said, no one is completely sure why sometimes the cool air sinks to form lacunosus, while other times the warm air rises in pockets to form the opposite arrangement of *cloudlets* with gaps between (such as the one on p14).

Top: over Norwich, Norfolk, UK, by Andrew Sidenis (Member 12532)
Bottom: over Montefiore dell'Aso, Italy, by Matteo Costagliola (Member 2656)

37

Either 'jet-stream Cirrus' or contrails left from an amazing Red Arrows fly-by.

☐ **I spotted radiatus**   Points scored: ⬜35 + ⬜30

Add to page 7

Date: . . . .   Time: . . . .

Location: . . . .

Weather conditions: . . . .

☐ Photographed   Image file(s): . . . .

Cloud-collecting Points:
Any radiatus .... 35 points ☐
Bonus for jet-stream Cirrus, or Cirrus radiatus, that extends overhead, from horizon to horizon .... 30 points ☐

Variety is found in: Cumulus (p9), Stratocumulus (p11), Altocumulus (p15), Altostratus (p17), Cirrus (p19).
Don't confuse with: undulatus (p35).
Clouds are often both varieties at once.

38

RADIATUS

WHEN a layer of cloud rolls or clumps extends in long lines that stretch off to the horizon, the effect of perspective makes these lines converge, like railway tracks, towards a point. Such a formation is a variety known as radiatus, and it can be found at all three *cloud levels*.

Cumulus radiatus, or 'cloud streets'.

Sometimes these lines of cloud form parallel to the wind direction, and sometimes perpendicular to it. They are more commonly perpendicular in low clouds, such as Cumulus (p9) or Stratocumulus (p11), when the long lines are known as 'cloud streets'. These tend to cause glider pilots to wet themselves with excitement, for they act as avenues of lifting air along which a glider can reliably gain altitude.

When high clouds are of the radiatus variety, the lines are often parallel to the wind. The most eye-catching examples appear when it is not just any old winds that cause them, but the 180mph winds of the jet streams, which encircle the globe in the mid-latitudes. Known as 'jet-stream Cirrus' (see opposite) these radiatus formations occur in clouds at the top of the *troposphere*. As the Cirrus (p19) is teased out by the high winds, it can occasionally seem to extend all the way from one horizon right overhead to the opposite one. The perspective causes the cloud rows to bulge dramatically between converging at each horizon. Such is a radiatus cloud of which any cloud collector would be proud, even though it is as good as impossible to photograph, because it takes up the whole sky.

Over The Bahamas, by Paul Cooper (Member 1523)

Differing wind directions reveal this Cirrostratus fibratus cloud to be in two layers.

☐ **I spotted duplicatus**   Points scored: 25 + 15   Add to page 7

Date: . . . .   Time: . . . .

Location: . . . .

Weather conditions: . . . .

☐ Photographed   Image file(s): . . . .

Cloud-collecting Points:
Any duplicatus .... 25 points ☐
Bonus for when the duplicatus variety
is revealed in Altocumulus by a long-
lasting sunrise or sunset .. 15 points ☐

Variety is found in: Stratocumulus (p11),
Altocumulus (p15), Altostratus (p17),
Cirrus (p19), Cirrostratus (p23).
Don't confuse with: two different genera
e.g. Cirrostratus (p23) over Stratus (p13).

40

Over Fochabers, Moray, Scotland, by Anne Burgess (Member 1481)

DUPLICATUS

SOME clouds earn substantial collecting points because they are rare, and others just because they are very difficult to identify. It is the latter case for the variety duplicatus, in which a layered cloud occurs at two altitudes at the same time.

The light and shade of a low Sun can reveal Altostratus duplicatus layers.

When duplicatus occurs in the low Stratocumulus cloud (p11), which contains water droplets rather than ice crystals, the lower of the two layers is usually too thick to see through, and so obscures the higher one. It is possible to notice that the cloud is divided into two layers only when the higher one appears through gaps in the lower. And, truth be told, in this low cloud, it's not a sight worth getting excited about. You might think the same could be said for duplicatus varieties in Altocumulus (p15) and Altostratus (p17), but then these mid-level clouds can be rather less opaque. So when the Sun is very low in the sky, the duplicatus formation can become visible as the lower of the two layers is darkened by the Earth's shadow, while the higher is bathed in ruby hues. In fact, Altocumulus duplicatus can produce the most gloriously prolonged sunsets.

In the high Cirrus (p19) and Cirrostratus (p23) clouds, you can most readily identify the split layers of duplicatus when they happen to have the filaments of the fibratus species (p31). The fibres of each layer can point different ways with differing wind directions at each altitude. Since these clouds are made of ice crystals and so are generally semi-transparent, the layers appear as one, with a beautiful cross-hatched pattern.

This is officially known as a 'fancy' pileus. (Actually, it's not, it's just a pileus.)

| ☐ **I spotted pileus** | **Points scored:** 45 + 15 |

Add to page 7

Date: .... Time: ....

Location: ....

Weather conditions: ....

☐ Photographed   Image file(s): ....

Cloud-collecting Points:
Any pileus .... 45 points ☐
Bonus for one that forms over the top
of a Cumulus congestus cloud that
looks like Donald Trump ... 15 points ☐

Normally found in the company of:
Cumulus congestus (p9),
Cumulonimbus (p27),
Don't confuse with: lenticularis (p29),
cap clouds (p59).

ACCESSORY CLOUDS & SUPPLEMENTARY FEATURES

Over Dubai, United Arab Emirates, by Sandra Malone (Member 14202)

PILEUS

THE most short-lived of all the *accessory clouds*, pileus is also the most beautiful. It shares much in common with lenticularis (p29) and cap clouds (p59), which form when a *stable airstream* rises to pass over raised ground. In the case of pileus, however, the obstacle is not rocky terrain, but something altogether more ephemeral – another cloud.

Pileus clouds are the comb-over hairstyle of the cloud world.

Pileus looks rather like a smooth, white beret, or, perhaps, a Donald-Trump, comb-over hairstyle. It is a horizontal cap cloud that appears momentarily on top of the crisp, cauliflower summit of a Cumulus congestus (p9), or the softer one of a young Cumulonimbus (p27). Pileus can appear as one of these large *convection clouds* develops upwards and encounters a moist stable airstream blowing above. This is forced to rise by the vigorous currents surging up the centre of the cloud below, cooling it just enough for some of its moisture to condense into droplets. These evaporate as the airflow sinks back down again past the convection cloud.

A pileus earns high points for the cloud collector because, unlike its relation, velum (p45), it never hangs around for long. Cloudspotters have to be sharp-eyed to add one to their collection. The vigorous convection cloud that made it inevitably continues its rise, pushing its bald head through the hairstyle. Donald Trump's head will eventually do the same.

43

Velum is the thin, horizontal strip of cloud in front of the Cumulus congestus.

| ☐ **I spotted velum** | **Points scored:** 15 + 10 | Add to page 7 |

Date: .... Time: ....

Location: ....

Weather conditions: ....

☐ Photographed   Image file(s): ....

Cloud-collecting Points:
Any velum .... 15 points ☐
Bonus for when you notice it still hanging around after the convection clouds have dissipated .... 10 points ☐

Normally found in the company of: a group of Cumulus congestus (p9) or Cumulonimbus (p27).
Don't confuse with: pileus (p43).

Over Teruel, Spain, by Luis Antonio Gil Pellin (Member 14194)

VELUM

VELUM is an *accessory cloud* that turns up in the same sort of places as the pileus cloud (p43). Though they form in a similar way, they have quite different natures. Unfortunately for velum, it is usually the less attractive and more ponderous of the pair.

Velum is Latin for ship's sail. This name is a tad misleading for it is a thin horizontal patch of cloud, rather than one that hangs downwards from a tall mast and catches the wind. You'll spot a velum cloud either just above, or around the sides of, a group of large *convection clouds,* such as Cumulus congestus (see p9) or Cumulonimbus (p27). Observing the clouds from a distance, the velum usually looks like a white or grey strip that can be separate from, or mixed in with, the convection clouds.

While the smaller pileus appears locally over individual convection clouds, velum is often spread over a very large area. If the cloud layer was not already around before the convection clouds grew up through it, the velum can sometimes have resulted from the tops of Cumulus clouds spreading out upon reaching a layer of *stable air* above. When, later, more powerful Cumulus clouds finally burst through the stable air layer, the velum remains loitering at their flanks for some time.

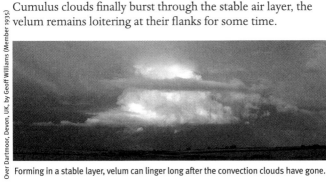

Over Dartmoor, Devon, UK, by Geoff Williams (Member 1935)

Forming in a stable layer, velum can linger long after the convection clouds have gone.

Is that a hoodie brandishing an alcopop? No, it's a pannus in moisture-laden air.

☐ **I spotted pannus**   Points scored: 10 + 10   Add to page 7

Date: ....   Time: ....

Location: ....

Weather conditions: ....

☐ Photographed   Image file(s): ....

Cloud-collecting Points:
Any pannus .... 10 points ☐
Bonus for correctly predicting to your
companion that it will rain within five
minutes, and being right .. 10 points ☐

Normally found in the company of:
Cumulus congestus (p9), Cumulonimbus
(p27), Nimbostratus (p25).
Don't confuse with: Cumulus fractus
(p9) beneath Altostratus (p17).

Over Mortimer Common, Berkshire, UK, by Stephen Burt (Member 2814)

PANNUS

DON'T get too excited about adding pannus to your cloud collection. When you do spot one, you're likely to be rather underwhelmed, for they aren't good-lookers.

Loitering in the saturated atmosphere just below rain clouds, they resemble some sort of cloud version of

An ominous sky is always made a little more ominous by pannus clouds.

hoodies, killing time outside McDonald's on a Saturday night. These dark shreds of cloud, strictly classified as Stratus fractus (see p13), give the sky a threatening air. The atmosphere below a precipitating cloud can become very humid, on account of all the moisture falling through it. Only the slightest rising gust can then cool the air enough for some of this moisture to condense into tiny droplets, which hang around as wisps of thin cloud.

If it is not raining or snowing when you notice dark shreds of pannus below a forbidding sky, you can be confident that it very soon will be. Pannus are the five-minute-precipitation-warning of the cloud world.

As with the kids on the High Street, the sinister appearance of pannus clouds owes a lot to their surroundings. The shreds of cloud need only be thick enough to block a little light for our eyes to register them as darker than the thick, dark rain clouds above. Away from their precipitous context, pannus would be seen for the weedy wisps that they are. The same could be said of the prepubescent 14-year-olds, once stripped of their mates and ubiquitous hoods.

Over Strathfield Mortimer, Berkshire, UK, by Stephen Burt (Member 2814)

47

As the underside of its anvil reveals, this Cumulonimbus needs milking.

☐ **I spotted mamma**	**Points scored:** 30 + 25	Add to page 7

Date: .... Time: ....

Location: ....

Weather conditions: ....

☐ Photographed   Image file(s): ....

Cloud-collecting Points:	**Normally found in the company of:**
Any mamma .... 30 points ☐	Stratocumulus (p11), Altocumulus (p15), Altostratus (p17), Cirrus (p19), Cirrocumulus (p21), Cumulonimbus (p27), contrails (p63), pyrocumulus (p65).
Bonus for when you're far enough away to see mamma under a Cumulonimbus anvil (pp27, 57) .... 25 points ☐	

MAMMA

'WHAT on Earth are those?' is the usual reaction when people see photographs of mamma clouds. Also known as 'mammatus', these *supplementary features* hang down from a layer of cloud in smooth or rough pouches that often have the appearance of udders (which is what 'mamma' means in Latin).

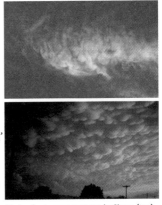

It's harder to spot mamma in Cirrus (*top*) than under a Cumulonimbus anvil (*bottom*).

With such an other-worldly, Independence-Day appearance, mamma are a must-have for any cloud collection. They can be found on a whole range of cloud types (see opposite) but the most dramatic examples occur on the underside of the huge anvils, known as incus (p57), that spread out at the top of mature Cumulonimbus storm clouds (p27) and can cover all the visible sky.

Some claim that mamma are harbingers of stormy weather, and what with the association between these pendulous cloud-boobs and Cumulonimbus, you might think they have a point. But mamma tend to form at the rear, rather than the front, of storms. Once you see mamma formations above you, the storm has usually passed over, or missed you entirely.

Each lobe of mamma is typically one to two miles across, and appears for around ten minutes. There are several theories about why they form, but an extensive 2006 scientific review of all the studies to date concluded that no one's really sure.

Virga is rain or snow that evaporates before reaching the ground.

| ☐ **I spotted virga** | Points scored: 25 + 15 | Add to page 7 |

Date: . . . . Time: . . . .

Location: . . . .

Weather conditions: . . . .

☐ Photographed   Image file(s): . . . .

Cloud-collecting Points:
Any Virga . . . . 25 points ☐
Bonus for when the trail of virga falls
through an abrupt change of wind and
develops a sharp corner ... 15 points ☐

Normally found in the company of:
Cumulus (p9), Stratocumulus (p11),
Altocumulus (p15), Altostratus (p17),
Cirrocumulus (p21), Nimbostratus (p25).
Don't confuse with: Cirrus (p19).

VIRGA

WHEN you look up to find jellyfish floating above, you are either diving or beneath the cloud *supplementary feature* known as virga.

In essence, this is just a cloud raining or snowing, but with one important difference: the precipitation never reaches the ground. If the droplets or ice crystals (or anything between the two) fall through air that is warm enough and/or dry enough, they can evaporate before ever landing.

When wind speeds vary considerably with height, virga can be very slanted (*top*) or have abrupt corners (*bottom*).

The appearance of virga from the ground is of trails that hang down like tentacles from clumps of cloud, waving not in the currents of the ocean, but in those of the lower atmosphere. When virga occur below low-level clouds, they are composed of water droplets, and appear grey. When they consist of ice crystals, having fallen from mid- or high-level clouds, they have a much paler appearance. But beware: this distinction is a tenuous one, because our eyes judge colour and tone relative to the brightness of the background. The same trail of virga can appear whiter or greyer depending on the sky behind. Fallstreak holes (p69) are specific cases of virga falling from a layer of *supercooled droplets* to leave a hole behind.

When a cloud's precipitation can be seen to reach all the way to the ground, it is no longer called virga, but 'praecipitatio'.

Bottom: over Acle Bridge, Norfolk, UK, by Jackie Clover (Member 11639)

51

ACCESSORY CLOUDS & SUPPLEMENTARY FEATURES

Arcus is like the front bumper of a speeding storm cloud.

☐ **I spotted arcus**   Points scored: 30 + 25   _{Add to page 7}

Date: .... Time: ....

Location: ....

Weather conditions: ....

☐ Photographed   Image file(s): ....

Cloud-collecting Points:
Any arcus .... 30 points ☐
Bonus for one that clearly protrudes
outwards in a ledge, which is known as
a shelf cloud .... 25 points ☐

Normally found with: large Cumulus
congestus (p9), Cumulonimbus (p27)
and multicell or supercell storms.
Don't confuse with: 'wall clouds' in the
rising air at the back end of storms.

52

Over Pewaukee, Wisconsin, US, by Nancy Wolck (Member 14205)

ARCUS

STORM chasers tend to have an abundance of arcus in their cloud collections, for this *supplementary feature* is rather like the front bumper of a storm cloud – a long, dark, horizontal roll or shelf running along the base of the storm cloud's front edge (around the registration plate). So arcus is the first cloud feature to arrive as the storm runs you over.

Top: An arcus. *Bottom:* The horizontally extended form, known as a shelf cloud.

Like those other brute-cloud groupies, tuba (p55) and incus (p57), arcus hang out only in the company of hefty Cumulus congestus (p9) or Cumulonimbus (p27) clouds or those most brutish of all cloud systems, the fierce *multicell* and *supercell storms*. Arcus forms as the cold air that is dragged down by all the precipitation falling within storm clouds splays outwards upon reaching the ground. As it spreads around the storm, it burrows under the warmer, less-dense air at ground level. This is lifted most forcefully in the direction of the cloud's movement, forming a 'gust front', in which the warmer air's moisture can condense into water droplets that appear as arcus.

Sometimes, arcus can protrude forwards from the storm as a dark, ragged ledge, known as a 'shelf cloud'. More rarely, the lifting motion can cause a wave of rising and falling air that races ahead of the storm, the arcus appearing as a roll cloud, which rotates within the wave as it travels (p67).

53

A waterspout is when a tuba touches down on an expanse of water.

| ☐ **I spotted tuba** | **Points scored:** 35 + 25 | Add to page 7 |

Date: . . . . Time: . . . .

Location: . . . .

Weather conditions: . . . .

☐ Photographed   Image file(s): . . . .

Cloud-collecting Points:
Any tuba . . . . 35 points ☐
Bonus for when the vortex of air becomes so vigorous that the tuba extends to the ground . . . . . . 25 points ☐

Normally found in the company of:
individual Cumulus congestus (p9) and Cumulonimbus (p27), but most prominently at the inflow of huge multicell and supercell storm systems.

54

THE air below a storm cloud is often a wild confusion of blustery, gusty and not-at-all-tranquil winds. But when the storm develops from a single Cumulonimbus cloud (p27) into a co-ordinated system, known as a *multicell* or *supercell storm*, the mêlée of air currents becomes much more organised. This is when a tuba can form.

The first tentative signs of a cloud finger.

Resembling a cloud finger descending from the storm's base, the tuba forms in the air sucked upwards into the storm to feed its vigorous vertical growth. Like an upside-down version of bath water going down a plug hole, the rising air can start rotating in a vortex. In a big storm cell, the rapidly rising air expands and cools enough for some of its moisture to condense to form the walls of the tuba. Also known as a 'funnel cloud', it can be the birth of a tornado.

A tuba can also form when the air is not rising but sinking from the base of individual clouds, such as Cumulus congestus (p9) and Cumulonimbus. Dragged towards the ground by the cloud's heavy showers, this sinking air can cause vortices to form. These are rarely as violent as the upward ones, so tubas are less pronounced. They herald not tornadoes, but the less ferocious landspouts or waterspouts.

Whatever a tuba is heralding, keep your distance when adding it to your cloud collection – just in case it has a mind to add a cloudspotter to its own collection of flying debris.

Over Castle Rock, Colorado, US, by Jim Karanik (Member 12842)

It's not Armageddon over Austin, Texas – just the incus of a Cumulonimbus.

☐ **I spotted incus**   **Points scored:** 20 + 15   Add to page 7

Date: .... Time: ....

Location: ....

Weather conditions: ....

☐ Photographed   Image file(s): ....

Cloud-collecting Points:
Any incus .... 20 points ☐
Bonus for when this canopy of ice crystals has a ribbed or streaked appearance .... 15 points ☐

Normally found in the company of:
Cumulonimbus (p27)
Don't confuse with: the mushroom cloud of a nuclear explosion.

56

Over Austin, Texas, US, by Ginnie Powell (Member 10927)

INCUS

AN incus is a part of a
Cumulonimbus cloud (p27).
In fact, it is the most distinctive
feature of the storm cloud: the
huge canopy of ice crystals that
spreads out at its top. Often
covering hundreds of square
miles, it is much larger than
any of the other *supplementary*

You have to be a long way away to be
able to see a storm cloud's incus.

features, and gives a mature storm cloud the distinctive shape
of a blacksmith's anvil (the meaning of incus in Latin).

Why does an incus form? One minute the cloud is growing
upwards, building from a Cumulus congestus tower (see p9),
its summit softening as the droplets there freeze into ice crystals, and the next it starts spreading outwards in all directions.
The vigorous vertical development of an enormous storm
cloud is blocked, and it is forced to splay outwards, when it
encounters what meteorologists call a *temperature inversion*.

On average, air becomes colder the higher you go through
the lower atmosphere. But only on average. Often, a situation
occurs when a region of warmer air blows over the top of a
colder one. Such an 'inversion' of the normal temperature
profile tends to act as an invisible ceiling to the growth of
clouds. All of a sudden, the warmer air within the cloud is no
longer warmer, and so less dense, than the air around. So it
stops floating upwards. The inversion that causes the top of a
Cumulonimbus to spread as incus is usually the one marking
the boundary between the *troposphere* and the *stratosphere*.
Called the *tropopause*, it's where the air temperature no longer
falls with altitude, and may even begin to rise.

Over Llanos de Gea, Teruel, Spain, by Luis Antonio Gil Pellín (Member 14194)

57

A cap cloud, known to some as a pink-fluffy-bobble-hat cloud.

☐ **I spotted a cap or banner cloud** Points scored: 30 + 20

Add to page 7

Date: .... Time: ....

Location: ....

Weather conditions: ....

☐ Photographed   Image file(s): ....

Cloud-collecting Points:
Cap or banner cloud .... 30 points ☐
Bonus for a cap in layers (above) or a
banner cloud showing turbulent whirls
(above, opposite)............... 20 points ☐

Typical altitudes: around, or just
above, a mountain summit.
Precipitation: from cap (but rare).
Don't confuse with: Pileus (p43),
Stratus (p13) or Stratocumulus (p11).

CAP & BANNER

WHILE cap and banner clouds form in slightly different ways, they share the distinction of being clouds that hang out around mountain summits.

When a cap cloud forms, it tends to look like a hat, perched upon the mountain's head. Sometimes it looks like a humble skullcap. Other times, it splays out in a full mother-in-law-at-a-wedding extravaganza. Occasionally, the mountain seems to be wearing one hat on top of another, which is surely a mountain fashion

Top: A banner cloud. *Bottom:* The mother-in-law-at-a-wedding cap cloud.

faux-pas. Whichever it is, a cap cloud forms as a *stable airstream* rises to pass over a peak, cooling as it does so. It is a particular example of a lenticularis *species* (p29), in which the cloud lies over the mountaintop, rather than downwind from it.

Banner clouds form in the same places as cap clouds, but look more as if the mountain is letting its hair flow in the wind, and they form in a slightly different way from cap clouds. As a stiff wind blows over a pronounced high peak, the air pressure drops slightly behind the peak. This can cool the air enough for its moisture to condense briefly into droplets or ice crystals.

Cloudspotters should be careful not to add the wrong cloud to their collections. Clouds such as Stratus (p13) or Stratocumulus (p11) clinging to mountaintops won't do. Only a jaunty cloud hat or a mane of mountain hair earn the points.

Top: over The Eiger, Switzerland, by Bob Peterson (Member 12047)
Bottom: over Mount Rainier, Washington, US, by Ryan Verwest (Member 13523)

Enormous breakers on a calm sea: Kelvin-Helmholtz clouds in thick Cirrus.

☐ **I spotted Kelvin-Helmholtz**   **Points scored:** 55 + 30   Add to page 7

Date: ....   Time: ....

Location: ....

Weather conditions: ....

☐ Photographed   Image file(s): ....

Cloud-collecting Points:
Any Kelvin-Helmholtz .... 55 points ☐
Bonus for when the Kelvin-Helmholtz
appears along the top of a
lenticularis cloud (p29) .... 30 points ☐

Typical altitudes: from ground level
(in fog) to 40,000ft.
Precipitation: influence is not significant.
Don't confuse with: regular waves of
undulatus (p35) that aren't breaking.

60

KELVIN-HELMHOLTZ

THE Kelvin-Helmholtz wave cloud scores the highest number of points of all the clouds. Looking just like enormous waves breaking on the shore, it is rare, fleeting and the favourite of cloudspotting surfers. A well defined Kelvin-Helmholtz is the crown jewel in many a cloud collection, for it requires the cloudspotter to be blessed with eagle-eyed sky awareness and sheer blind luck. In one spotting alone, this cloud can help observers overtake their fiercest cloud-collecting rivals.

The breaking waves can appear in Cirrus (*top*) and Stratocumulus (*bottom*).

It appears at all three *cloud levels,* and can be thought of as a very specific example of the undulatus cloud *variety* (p35), tending to be found in Stratocumulus (p11), Altocumulus (p15) or Cirrus (p19) clouds. It can also sometimes be seen along the top edge of a layer of fog (p73). In all cases, the formation lasts no more than a minute or two.

The distinctive breaking-wave shape is caused by wind shear. When a patch of cloud forms at the boundary between a colder layer of air below and a warmer one above, and the upper layer is moving more rapidly than the lower one, the shearing between the two can make undulations appear in the cloud. If the difference in wind speeds is just right, the tops of the undulations are pushed ahead of the bottoms to form vortices that 'break' just like the ocean surf, dude.

Top: over Monument, Colorado, US, by Terry Robinson (Member 7004)
Bottom: over Gent, Belgium, by Frits Kuitenbrouwer (Member 13684)

Aircraft contrails – a nightmare for anyone shooting a period drama outside.

☐ **I spotted a contrail**   Points scored: 10 + 10   Add to page 7

Date: . . . .   Time: . . . .

Location: . . . .

Weather conditions: . . . .

☐ Photographed   Image file(s): . . . .

Cloud-collecting Points:
Any contrail .... 10 points ☐
Bonus for when the plane cuts a gap out of an existing layer of cloud to form a distrail ....10 points ☐

Typical altitudes: 28,000–40,000ft.
Precipitation: none.
Don't confuse with: too straight to be mistaken for other clouds.

CONTRAILS

BEFORE the start of the First World War and the advent of high-altitude flight, our skies appeared very different from the way they do today – there were no condensation trails, or contrails, which form in the exhaust of aircraft.

A distrail, short for dissipation trail.

There's no confusing these man-made clouds with the natural ones. Following the aircraft's path, contrails tend to appear as long, straight slashes of white across the blue. In the vicinity of airports, however, they can sometimes form large loops, due to the stacking formation of aircraft waiting to land.

The length of time contrails remain in the sky – or indeed whether they form at all – varies greatly depending on the air conditions up at cruising altitude. When it's cold enough and moist enough, the *water vapour* contained in the plane's hot exhaust gases mixes with the very cold air to condense and form ice crystals. In some conditions, these soon evaporate. In others, they can persist for hours, the ice crystals absorbing water vapour from the surrounding air to grow in size and spread out in the high winds. In this way, contrails often encourage the formation of Cirrus (p19), Cirrocumulus (p21) and Cirrostratus (p23) ice-crystal clouds.

Bonus cloud-collecting points are awarded for spotting a distrail. The opposite of a contrail, this is when a plane cuts out a gap as it flies through a cloud layer. This happens when the heat and turbulence of its engine make the cloud droplets evaporate, or when it introduces *icing nuclei* that encourage the cloud's *supercooled droplets* to freeze and fall below.

63

Man-made pyrocumulus, also known as 'fumulus'.

☐ **I spotted pyrocumulus**   **Points scored:** 10 + 5

Add to page 7

Date: ....   Time: ....

Location: ....

Weather conditions: ....

☐ Photographed   Image file(s): ....

Cloud-collecting Points:	**Typical altitudes:** 0–5,000ft.
Any pyrocumulus .... 10 points ☐	**Precipitation:** rain possible if grows very large over volcanoes/forest fires.
Bonus for one that is so vigorous it leads to thunder and lightning .... 5 points ☐	**Don't confuse with:** smoke or volcanic ash alone, without any cloud.

PYROCUMULUS

WHEN clouds form in columns of air floating up from forest fires, they are known as pyrocumulus. Anything that is hot enough to produce strong *convection currents* can give rise to clouds if there is enough moisture around. Fierce forest fires and large volcanic eruptions can lead to pyrocumulus clouds that are large enough to produce lightning or tuba cloud features (p55), which lead to landspouts or waterspouts. Pyrocumulus also form over the towers of power stations, when they are sometimes known as 'fumulus'. This is one of the few man-made clouds – another being contrails (p63).

Forest fires, volcanoes and power stations not only provide the heat needed to form these clouds; they also introduce countless microscopic particles into the air that act as *condensation nuclei*, on which the cloud's droplets can form.

This means pyrocumulus droplets tend to be very small and plentiful, making the cloud appear thick – especially when mixed with ash or smoke.

Forest-fire smoke can have a reddish tinge, while pyrocumulus clouds can look dark grey (*right*) or bright white (*below*).

65

The queen of roll clouds is Australia's 'Morning Glory'.

| ☐ **I spotted a roll cloud** | Points scored: 30 + 10 | Add to page 7 |

Date: .... Time: ....

Location: ....

Weather conditions: ....

☐ Photographed Image file(s): ....

Cloud-collecting Points:
Any roll cloud .... 30 points ☐
Bonus for when the roll cloud passes right over you, and you stop what you are doing to watch it .... 10 points ☐

Typical altitudes: 1,000–5,000ft.
Precipitation: none itself, but it can be associated with storms.
Don't confuse with: other forms of arcus (p53) attached to storm clouds.

66

Over Gulf Savannah coast of Northern Queensland, Australia, by Gavin Pretor-Pinney (Member 1)

ROLL CLOUD

THIS is a long, low tube of cloud, which can appear to extend horizontally from horizon to horizon. Sometimes a roll cloud has a very smooth, silky surface. At other times, it can appear quite rough and bumpy. Roll clouds can move at speeds of up to 35mph, with the tube appearing to rotate as it rolls along. The direction of rotation is not as it would be for a solid tube rolling along the ground. In fact, the roll cloud rotates against its direction of travel – the cloud surface lifting at the front and dropping down at the back.

One famous roll cloud, 'Morning Glory', appears in Northern Queensland, Australia. This forms in a solitary wave of air and is caused by colliding sea breezes over the Cape York Peninsula. But most roll clouds are types of arcus (p53), and caused by storms. As the storm dissipates, gusting winds of cold air can continue to spread out ahead of it, and form a roll of cloud that separates away from the rest of the storm.

Bottom: over Collinsville, Oklahoma, US, by Jodie Justice (Member 14210)

Top and bottom: Roll clouds can move ahead of big storms that are dissipating.

67

Once some droplets freeze, they rob others of moisture and start to descend.

☐ **I spotted a fallstreak hole**   Points scored: 35 + 15   Add to page 7

Date: ....   Time: ....

Location: ....

Weather conditions: ....

☐ Photographed   Image file(s): ....

Cloud-collecting Points:
Any fallstreak hole .... 35 points ☐
Bonus for when the trail of ice crystals
has iridescent colours (p81) as the Sun
shines through it .... 15 points ☐

Typical altitudes: 6,500–20,000ft.
Precipitation: the crystals evaporate
before reaching the ground.
Don't confuse with: Cirrus (p19).

FALLSTREAK HOLE

THEY look bizarre, but fallstreak holes are not actually that rare. Also called 'hole-punch clouds', they are crisp gaps in mid- or high-level cloud layers, below which dangle trails of ice crystals.

A cigar-shaped fallstreak hole, caused by a plane climbing through the cloud.

To form a fallstreak hole, the cloud layer must consist of *supercooled droplets* – when its water is in liquid form despite temperatures at cloud level being well below 0°C. This is actually quite common, for pure water suspended as droplets in the air behaves very differently from tap water in the freezer. If there aren't enough of the right sort of tiny particles in the atmosphere to act as *icing nuclei*, on to which they can start to freeze, droplets remain liquid until temperatures drop to around –40°C. They 'want' to freeze, but can only do so when there are seeds on which the crystals can begin to grow.

A fallstreak hole forms when one region of the cloud finally starts to freeze and begins a chain reaction. All the moisture from the supercooled droplets in the area rushes to join the ice crystals, which quickly grow big enough to fall below. A form of virga (p51), the trail of ice crystals doesn't tend to reach the ground, but evaporates before getting that far.

What starts the freezing? Sometimes it's ice crystals falling into the cloud's droplets from a higher Cirrus cloud (p19). At other times, it is caused by the exhaust of a plane climbing or descending through the cloud to form a 'distrail' (see p63). Particles in the exhaust act as icing nuclei that start the freezing.

An upside-down horseshoe, considered unlucky by some, but never cloudspotters.

☐ **I spotted a horseshoe vortex**   Points scored: 50 + 20   Add to page 7

Date: .... Time: ....

Location: ....

Weather conditions: ....

☐ Photographed   Image file(s): ....

Cloud-collecting Points:
Any horseshoe vortex ...... 50 points ☐
Bonus for managing to catch it on film, so that you can see the way the winds make it rotate .... 20 points ☐

Typical altitudes: 2,000–5,000ft.
Precipitation: not from this snippet, but it does often form in the vicinity of large storm clouds.
Don't confuse with: unmistakeable!

70

Over Vienna, Austria, by Jo Gardner (Member 340)

OTHER CLOUDS

HORSESHOE VORTEX

WHAT a subtle little wisp of cloud the horseshoe vortex is! It is easily missed by anyone other than the most keen-eyed cloudspotter, intent on adding it to their collection. The rare and fleeting horseshoe vortex cloud appears for just a minute or so before evaporating. Anyone lucky enough to spot one must take a photo if they want to be believed by their cloud-collecting friends.

Had the French named it, this would surely be called the 'croissant cloud'.

This cloud forms in a region of rotating air, or vortex. While the familiar orientation for a vortex is vertical (see tuba, p55), they can occasionally develop on a horizontal axis. This is when the gently rotating crescent of the horseshoe vortex cloud can form. The movement of air seems to result from an updraught that is sent into a spin when it reaches a sudden change in the horizontal winds above. Rarely are conditions right for a cloud to appear within the vortex but, when they are, the air in the upper arc of the vortex cools enough to develop a rotating crescent of cloud. One of the best places to spot horseshoe vortex clouds is in the vicinity of *supercell storms*. The winds rushing in to feed the storm's growth can lead to just the right sort of shearing air currents.

This beautiful little cloud may not lead to any precipitation, but it will rain down luck upon anyone fortunate enough to spot it – as well as impressive cloud-collecting points.

Top: over Jasper, Alberta, Canada, by Eric Kenwald (Member 14215)
Bottom: over Bicheno, Tasmania, Australia, by Joseph Murray (Member 14433)

Radiation fog forms on long, cold, damp nights, below cloud-free skies.

| ☐ **I spotted fog or mist** | Points scored: 15 + 5 | Add to page 7 |

Date: .... Time: ....

Location: ....

Weather conditions: ....

☐ Photographed   Image file(s): ....

Cloud-collecting Points:
Any fog or mist .... 15 points ☐
Bonus for crepuscular rays (p85)
appearing in the fog as the sunlight
begins to cast shadows ...... 5 points ☐

Typical altitudes: ground/sea level.
Precipitation: 'fog drip' moisture when droplets collide with solid surfaces.
Don't confuse with: airborne clouds, if you are the pilot of a plane.

FOG & MIST

THE distinction between fog and mist relates to visibility. Officially, you can see 1–2km in mist but no more than 1km in fog – one's just a thicker version of the other. Though fog is sometimes described as ground-level Stratus cloud (p13), since that's the lowest of the main clouds, it often forms quite differently.

Fog appears if air is cooled enough by its proximity to the ground or water surface for its moisture to condense into droplets. There are two main ways this cooling can happen.

'Radiation fog' forms after long, cold and clear nights. With no blanket of cloud cover to keep the warmth in, the ground quickly radiates the day's warmth into the night sky, and can cool the air enough to form droplets. On higher ground, the cold, foggy air can sink downhill and gather as 'valley fog'.

'Advection fog' occurs when air cools as it blows over a warmer surface to a colder one. If these are ocean surfaces, it's called 'sea fog'. Then there's 'steam fog' – when cold air blows over warmer water, such as a lake, and the *water vapour* that evaporates off the surface instantly cools to form droplets.

That's not the end of it. There's also 'upslope fog', 'hill fog', 'ice fog' (see p75), 'haar' and 'frontal fog'. No matter which one it is, cloudspotters will never get closer to a cloud than when they're enveloped in fog or mist.

Left: Steam fog can form over a lake.

Right: Valley fog seen from the hillside.

Halo phenomena can appear as the sun shines through diamond dust.

☐ **I spotted diamond dust**   Points scored: 30 + 15   Add to page 7

Date: .... Time: ....

Location: ....

Weather conditions: ....

☐ Photographed   Image file(s): ....

Cloud-collecting Points:	**Typical altitudes:** ground level.
Any diamond dust .... 30 points ☐	**Precipitation:** the tiny ice crystals
Bonus for when it scatters the	are generally too few to gather
sunlight to form a halo phenomenon	significantly on the ground.
(see pp93–99) .... 15 points ☐	**Don't confuse with:** magic pixie dust.

74

DIAMOND DUST

THE sight of diamond dust glittering in the sunlight is unforgettable. A kind of fog made of ice crystals, it is often not thick enough to reduce visibility and its presence is then revealed only by the way the crystals glint in the light as they tumble through the air.

Diamond dust is sometimes also known as 'ice fog', but this term tends to refer to a thicker ice-crystal fog that

Crystals of diamond dust sparkle as they tumble through the night air.

reduces visibility and consists of less regular crystal shapes. For classic diamond dust, temperatures need to be lower than about –20°C. This means that the air's *water vapour* tends to collect directly as floating ice crystals, rather than condensing into droplets, which then freeze. The result is a ground-level version of a Cirrus (p19) or Cirrostratus (p23) cloud.

Such conditions are common in polar regions, particularly in the Antarctic, where the crystals grow very slowly, leading to diamond dust of regularly shaped little ice prisms. These do more than just glisten. By reflecting and *refracting* the light waves passing through, the crystals of diamond dust can produce extremely pure and extensive arcs, spots and rings of light, known as *halo phenomena* (see pp93–99).

But cloudspotters needn't go to Antarctica to add these shimmering crystals to their cloud collection. Pure diamond dust, producing glorious halo phenomena, can also form downwind from ski-resort snow machines.

Due to their height, nacreous can shine for as long as two hours after sunset.

| ☐ **I spotted nacreous** | Points scored: 45 + 20 |

Add to page 7

Date: .... Time: ....

Location: ....

Weather conditions: ....

☐ Photographed   Image file(s): ....

Cloud-collecting Points:
Any nacreous .... 45 points ☐
Bonus for watching through the sunset
until it is no longer lit, and the colours
suddenly switch off .... 20 points ☐

Typical altitudes: 10–20 miles.
Precipitation: none.
Don't confuse with: iridescence (p81)
that appears in much lower clouds
within the troposphere.

NACREOUS

FORMING 10–20 miles up, in the *stratosphere*, at −85°C, nacreous clouds show beautiful iridescent pastel hues as they scatter the light from the Sun when it is just below the horizon.

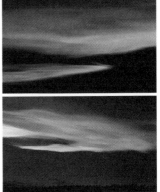

Sometimes called 'mother-of-pearl clouds', their tiny, uniform ice crystals are very good at *diffracting* sunlight. This separates the light into bands of colour, to create a much more dramatic version of the iridescence (p81) some-times seen in lower clouds.

The hues of nacreous clouds change with the Sun's angle over the horizon.

Also known as 'polar stratospheric clouds' since they tend to appear over higher-latitude regions of the world, nacreous clouds are like a stratospheric version of the lenticularis *species* of wave cloud (p29). They form when the atmosphere is so *stable* that waves produced as air flows over mountains down at ground level are transferred up through the atmosphere, and push moisture into the lower stratosphere. The best time of year to spot them is in winter, when temperatures are lowest.

Sadly, these most mesmerising of clouds are also the most destructive for our environment. Their tiny ice crystals act as catalysts that speed up the destruction of the protective ozone layer by the CFC gases we've released into the atmosphere. For clouds to have such other-worldly beauty, there always had to be a catch.

77

Noctilucent clouds at the fringes of space, shot from a plane flying at 41,000ft.

Over the Atlantic, south of Iceland, by Bill Valentine (Member 14,211)

☐ **I spotted noctilucent**	**Points scored:** 45 + 20	Add to page 7

Date: . . . . Time: . . . .

Location: . . . .

Weather conditions: . . . .

☐ Photographed   Image file(s): . . . .

Cloud-collecting Points:
Any noctilucent . . . . 45 points ☐
Bonus if they're at latitudes below 50°
(south of Britain, north of the southern
tip of New Zealand) . . . . 20 points ☐

Typical altitudes: 30–50 miles.
Precipitation: from 50 miles up? No.
Don't confuse with: Cirrostratus (p23).
No higher than 6 miles up, these are in
shadow while noctilucent are still bright.

NOCTILUCENT

THE mysterious noctilucent clouds are higher than any other cloud in the atmosphere. Also known as 'polar mesospheric clouds', they have an eerie, bluish-white appearance, often showing delicate ripples or billows.

Ripples make the cloud look thicker and thinner when viewed from an angle.

Noctilucent clouds form in the *mesosphere*, at altitudes of 30–50 miles – almost at the limit of the atmosphere. Being so high means that, in the higher latitudes, where they are most frequently seen, noctilucent clouds shine out against the night sky well after the Sun has dropped over the horizon. They still catch the sunlight when the rest of the sky is dark. Their name comes from Latin for 'night shining'.

Quite how noctilucent clouds form is by no means clear. The mesosphere is a region where air temperatures can be as low as –125°C but there is very little moisture at all. No one knows why the ice crystals that make up this cloud arise in such a dry and remote part of the atmosphere.

Historically, noctilucent clouds have tended to be spotted at latitudes higher than 50° during the summer months. It now seems that they are appearing over much larger regions of the world and more frequently. Some scientists have speculated that this change might be related to global warming.

The best times for cloudspotters to try to add noctilucent clouds to their collections is soon after sunset or before sunrise from May to August in the Northern Hemisphere, and November to March in the Southern Hemisphere.

Iridescence can be produced by ice crystals, like those of this Cirrocumulus.

Add to page 7

☐ **I spotted iridescence** Points scored: 20 + 10

Date: .... Time: ....

Location: ....

Weather conditions: ....

☐ Photographed   Image file(s): ....

Cloud-collecting Points:
Iridescence in any cloud except
nacreous (p77) .... 20 points ☐
Bonus for when it gives a cloud a
multicoloured lining ....10 points ☐

Seen in: thin Stratocumulus (p11),
Altostratus (p17), Altocumulus (p15),
Cirrocumulus (p21), lenticularis (p29)
and pileus (p43) edges.
Don't confuse with: glory.

IRIDESCENCE

IRIDESCENCE is the beautiful effect of bands of pastel colours that can appear when sunlight or moonlight passes through thin cloud. Also known as 'irisation', these mother-of-pearl colours are caused by the light being *diffracted* as it passes through the cloud. It is the same process that gives rise to coloured discs around the Sun or Moon, called coronae (p83). The light waves are dispersed as they pass around the cloud's tiny

Iridescence formed by cloud droplets in Altocumulus (*top*) and pileus (*bottom*).

droplets or ice crystals, with different wavelengths being spread out by different amounts. Each wavelength can also produce an interference pattern of light and dark fringes. All this means that the sunlight is separated into alternating fringes of colour.

Any thin cloud with droplets or ice crystals small and uniform enough can produce iridescent colours. Best viewed through sunglasses, the bands of pastel hues are reminiscent of those on oil slicks, and can be seen over thin Stratocumulus (p11), Altocumulus (p15), Cirrocumulus (p21) and Altostratus (p17). Nacreous clouds (p77) in the *stratosphere*, high above most weather clouds, show the most intense iridescence of all.

The colours can also appear at the edges of clouds thick enough to block much of the sunlight, such as lenticularis (p29). To claim that every cloud has a silver lining is therefore wrong – some have tutti-frutti coloured ones.

81

The width of a cloud's corona depends on the size of its droplets.

☐ I spotted a corona	Points scored: 20 + 15

Add to page 7

Date: .... Time: ....

Location: ....

Weather conditions: ....

☐ Photographed   Image file(s): ....

Cloud-collecting Points:
Any corona .... 20 points ☐
Bonus for when the colours are
distinct enough to extend from
red through to red .... 15 points ☐

Seen in: thin Stratocumulus (p11),
Stratus (p13), Altostratus (p17),
Altocumulus (p15), Cirrostratus (p23),
Cirrocumulus (p21).
Don't confuse with: glory (p87).

82

Over North Carolina, US, by Ally Summers (Member 56542)

CORONA

LOOK towards the Sun shining through thin cloud and you might find that it is surrounded by a corona. This is a bluish-white disc of light with a ruddy outer edge, often surrounded by rings of iridescent colours. Cloudspotters should be careful to protect their eyes by blocking the Sun with a hand. Coronae can be seen less painfully when clouds drift in front of a bright moon.

Cirrostratus corona around the Sun (*top*) and the Moon (*bottom*).

Closely related to cloud iridescence (p81), coronae are caused when the light is *diffracted* as it passes around a cloud's particles. Only if these are all very small and the cloud layer is thin will the colours of the corona appear distinct around the central bright disc. The smaller the cloud droplets, the larger the corona.

Cloudspotters should take care not to confuse a corona with a 22° halo (p93). Not only is the corona much smaller (the outer edge usually being less than 5° from the Sun or Moon – the width of three fingers, at arm's length), it also has a bright central disc, or 'aureole', while the halo is just a ring of light. Nor should they confuse it with a glory (p87), which appears in the opposite direction, looking away from the Sun.

Coronae can also be seen around car headlights viewed through a windscreen misted up with condensation. Anyone claiming cloud-collecting points for this is a terrible cheat.

83

'Jacob's ladder' occurs when crepuscular rays shine through holes in Stratocumulus.

☐ **I spotted crepuscular rays**   **Points scored:** 25 + 20   Add to page 7

Date: ....   Time: ....

Location: ....

Weather conditions: ....

☐ Photographed   Image file(s): ....

Cloud-collecting Points:
Any crepuscular rays .... 25 points ☐
Bonus for the much less obvious anti-crepuscular rays, which radiate from a point opposite the Sun ..... 20 points ☐

Seen in: (typically) Cumulus (p9) and Stratocumulus (p11), when water, dust and pollen particles scatter sunlight to show the cloud's shadow in 3D.

Over west coast of Levkas, Greece, by Angela Craggs (Member 4522)

CREPUSCULAR RAYS

EVEN if they don't know the name, most cloudspotters will have plenty of opportunities to add crepuscular rays to their collection of cloud optical effects. They're the familiar sunbeams that appear to burst from behind a Cumulus cloud (p9), or shine down through a hole in a Stratocumulus (p11).

Crepuscular rays appear when the path of sunlight is made visible by tiny atmospheric particles, too scarce to appear as cloud, but plentiful enough to scatter the light noticeably. Like fingers through a torch beam, the cloud shadows give edges to the rays. In spite of being almost parallel, these rays seem to radiate outwards from behind the cloud. This is just the same perspective effect as railway tracks seeming to widen as they get nearer.

Whenever you notice crepuscular rays from a low Sun, look to the opposite horizon for the far less obvious 'anti-crepuscular rays'. Appearing to emanate from a point directly opposite the Sun, these are the shadows cast by clouds behind you, like the shadow of someone shuffling behind you in a dusty cinema. Perspective makes them appear to converge in the distance. Few people ever notice anti-crepuscular rays – except vampire cloudspotters, eager for the arrival of night.

Top: Rays fanning out from behind a Cumulus. *Bottom:* Anti-crepuscular rays.

Top: over Death Adnach, Glenside, Scotland, by John MacPherson (Member 10504). Bottom: over Sulphur Springs Valley, Arizona, US, by John Annesley (Member 14212)

85

The Brocken spectre is a ghostly apparition caused by your own shadow.

| ☐ **I spotted a glory** | Points scored: 30 + 20 |

Add to page 7

Date: . . . .   Time: . . . .

Location: . . . .

Weather conditions: . . . .

☐ Photographed   Image file(s): . . . .

Cloud-collecting Points:
Any glory . . . . 30 points ☐
Bonus for all the effort of climbing a mountain and seeing a Brocken spectre (above) . . . . 20 points ☐

Seen in: Stratus (p13), Altocumulus (p15), Altostratus (p17), fog (p73).
Don't confuse with: a corona (p83), which has similar colours but is a disc around the Sun, not your shadow.

86

GLORY

CLOUDSPOTTERS must gain some elevation to add a glory to their collection of cloud optical effects, for this striking phenomenon is seen only with the Sun directly behind you, as it casts your shadow on to a layer of cloud. The glory, which looks like a series of rainbow rings around the shadow, is produced by cloud droplets reflecting, *refracting* and *diffracting* sunlight, although the exact optics are still not fully understood.

One of the easiest places to spot a glory is from that great cloudspotting location, the window of a plane. It can sometimes appear around the plane's shadow, cast on to a nearby layer of cloud or fog. When the cloud's some distance away, the shadow is absent and you just see the coloured rings.

Bonus cloud-collecting point are awarded for the most eerie form of glory – the 'Brocken spectre'. This is when the rings appear around your own shadow as you look at cloud from a mountain ridge. The perspective can make the legs of your shadow flare out so, what with the multicoloured halo, it looks like a ghost from the 1970s.

Right: A glory around the shadow of a hang glider, flying over low cloud. *Below:* One seen through the fog from a bridge.

Top: over coast near Sydney, NSW, Australia, by Robert Seckold (Member 14208)
Bottom: from The Golden Gate Bridge, San Francisco, US, by Mila Zinkova (Member 11067)

The best rainbows are produced by large raindrops, a few millimetres across.

☐ **I spotted a rainbow**   Points scored: 15 + 15   Add to page 7

Date: ....   Time: ....

Location: ....

Weather conditions: ....

☐ Photographed   Image file(s): ....

Cloud-collecting Points:	**Seen in:** any rainfall in direct sun –
Any rainbow .... 15 points ☐	often from Cumulus congestus (p9) and Cumulonimbus (p27).
Bonus for a secondary bow, and for noticing Alexander's dark band between the two .... 15 points ☐	**Don't confuse with:** Circumzenithal arc (p97), halo phenomena (pp93–99).

88

RAINBOW

WE all love rainbows, but don't expect to earn many cloud-collecting points for seeing one. They're just too easy to notice. How ironic that rainbows, appearing on average about ten times a year,* are actually much less frequent than *halo phenomena* (see pp93–99), which most people never notice.

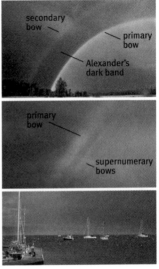

To see a rainbow, look towards a rain shower with your back to the Sun, which must be no higher than 42° above the horizon.** Passing into each raindrop and reflecting off the back inner surface, the sunlight is *refracted* as it enters and leaves. The paths of its constituent wavelengths are bent by different amounts, separating

Top and Middle: Primary, secondary and supernumerary bows, and Alexander's dark band. *Bottom:* When the sun is high, only the tip of the bow is visible.

out the colours. *Convection clouds* are the best sort for making rainbows, as they're more likely to produce showers when the sky around is clear, allowing direct sunlight to shine on them.

Besides the primary bow, a larger, fainter secondary bow can appear – the darker sky between the two being known as 'Alexander's dark band'. Within the primary bow, there are sometimes faint coloured fringes, called 'supernumerary bows'.

Bottom: over Mylor Harbour, Falmouth, Cornwall, UK, by Linda Bennett (Member 3270)

* See note about the frequency of optical phenomena on p112.
** Unless you are looking down, from a high building, say.

A cloudbow often appears at the same time as a glory (see bottom right of image).

| ☐ | **I spotted a cloudbow/fogbow** | Points scored: | 35 | + | 5 | Add to page 7 |

Date: .... Time: ....

Location: ....

Weather conditions: ....

☐ Photographed   Image file(s): ....

Cloud-collecting Points:
Any cloudbow or fogbow .. 35 points ☐
Bonus for managing to persuade
someone that it actually is the ghost of
a deceased rainbow .... 5 points ☐

Seen in: Stratus (p13), Stratocumulus (p11), Altocumulus (p15), Altostratus (p17), fog/mist (p73).
Don't confuse with: rainbow (p89), glory (p87).

Near Tenerife, Canary Islands, Spain, by Stefan Binnewies and Josef Pöpsel (Members 14206 and 14207)

CLOUDBOW

CLOUDBOWS are like rainbows, but with much paler colours. In fact, they often don't show any discernible colours at all – looking like albino rainbows, or the ghosts of rainbows past. They appear as the sunlight is reflected and *refracted* by the tiny droplets that make up low and mid-level clouds, rather than the much larger raindrops that produce rainbows (p89).

In order to see a cloudbow, you have to be looking towards cloud droplets on to which the Sun is shining from directly behind your line of vision. Such a viewpoint is possible only from above clouds, which is why cloudbows are usually seen from a plane, or a mountaintop. These are the same conditions in which to see the multicoloured ring around your shadow, called a glory (p87). If you see one, look for the other.

Cloudspotters wanting to add an albino rainbow to their collection, without taking to the air, should seek 'fogbows'. These are exactly the same, but appear in fog or mist (p73).

Cloud/fogbow colours are pale or absent because, at less than 0.1mm across, cloud droplets *diffract* sunlight more efficiently than larger raindrops, so the colours overlap more than those of a rainbow.

Look for fogbows when the sun shines brightly through thin fog or mist.

Twinkles from countless ice crystals of Cirrostratus add up to make a 22° halo.

| ☐ **I spotted a 22° halo** | **Points scored:** 25 + 20 | Add to page 7 |

Date: .... Time: ....

Location: ....

Weather conditions: ....

☐ Photographed   Image file(s): ....

Cloud-collecting Points:
Any 22° halo .... 25 points ☐
Bonus for a lunar 22° halo. This can appear round a full Moon, but is usually too dim to show colour .... 20 points ☐

Seen in: Cirrostratus (p23), Cirrus (p19), Cirrocumulus (p21), diamond dust (p75).
Don't confuse with: the much smaller coloured disc of a corona (p83).

92

Over Auckland, New Zealand, by Elizabeth Williams (Member 16232)

22° HALO

THE 22° halo is the most frequent of the many *halo phenomena* that can appear as sunlight is *refracted* through the ice crystals of thin layers of high clouds, such as Cirrus (p19), Cirrostratus (p23) and Cirrocumulus (p21), or the ground-level ice-crystal cloud, diamond dust (p75). Appearing

The inner edge often has a reddish tinge.

on about 100 days of the year,* the 22° halo looks like a large ring around the Sun or Moon. Its inner edge generally has a reddish tinge to it, with the region of the sky between halo and Sun appearing darker than that just outside the halo. When cloud cover is less than extensive, only parts of the halo appear.

Cloudspotters observing a 22° halo for the first time will be surprised at how much larger it appears compared with photographs (wide-angle lenses are invariably used to make it fit the frame). The distance from Sun or Moon to the edge of the ring is equivalent to the outstretched span of a hand held up at arm's length.

Though it's worth 30 cloud-collecting points, a 22° halo is not worth screwing up your eyesight for, so take note of the halo-spotting safety advice on p112.

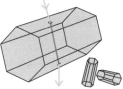

The **22° halo** can appear when a cloud's ice crystals have the shape of hexagonal columns. When these tumble in all directions, they refract the light most by an angle of 22°. Their combined sparkles form a ring of brighter light that appears as a halo around the Sun. To an observer, the angle between the Sun and the ring is 22°.

* See note about frequency on p112.

A sundog has a reddish edge towards the Sun and a bluish-white tail away from it.

| ☐ **I spotted sundogs** | Points scored: 20 + 15 |

Add to page 7

Date: .... Time: ....

Location: ....

Weather conditions: ....

☐ Photographed   Image file(s): ....

Cloud-collecting Points:
One sundog .... 20 points ☐
Bonus for when sundogs
appear on both sides of the
Sun at the same time ....15 points ☐

Seen in: thin Cirrostratus (p23), Cirrus (p19), Cirrocumulus (p21), diamond dust (p75), contrails (p63).
Sun's elevation: below 40° from horizon.

SUNDOGS

ALSO known as 'mock suns' and 'parhelia', sundogs are large spots of light that can appear on one or both sides of the Sun, and level with it. The second most frequent *halo phenomenon* after the 22° halo (p93), sundogs appear on about 70 days of the year.* They are formed as sunlight is *refracted* through the ice crystals of thin layers of high clouds, such as Cirrus (p19), Cirrostratus (p23) and Cirrocumulus (p21), or those of the ground-level ice-crystal cloud, diamond dust (p75). The distance from the Sun of both spots of light is equivalent to the outstretched span of a hand held up at arm's length. Sundogs are brightest when the Sun is low, and visible only when it's below about 40° from the horizon. If the cloud isn't in the right parts of the sky, just one sundog appears.

Cloudspotters should learn to recognise the sky that tends to produce sundogs, as well as the other halo phenomena. Scan the blue for light effects when it first pales with the subtle veil of ice-crystal clouds.** Once these are white enough to be noticed by the riffraff, they're generally too thick for sundogs.

* See note about frequency on p112.
** See p112 for halo-spotting safety advice.

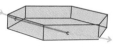

Sundogs can appear when a cloud's ice crystals are shaped like hexagonal plates and aligned almost horizontally (like falling leaves). When the Sun is very low in the sky, the crystals refract most light by 22°, so observers see their collective sparkles as bright spots on either side of the Sun.

Sundogs always appear level with the Sun.

Over Boston Docks, Lincolnshire, UK, by Nigel Counihan (Member 8777)

95

A circumzenithal arc, high in the sky, formed by diamond dust.

☐ **I spotted a circumzenithal arc**   **Points scored:** 45 + 25   Add to page 7

Date: ....   Time: ....

Location: ....

Weather conditions: ....

☐ Photographed   Image file(s): ....

Cloud-collecting Points:	**Seen in:** thin Cirrostratus (p23),
Any circumzenithal arc ..... 45 points ☐	Cirrus (p19), Cirrocumulus (p21),
Bonus for when it appears	diamond dust (p75).
at the same time	**Sun's elevation:** below 30° from the
as sundogs (p95) .... 25 points ☐	horizon, but best when at 22°.

96

CIRCUMZENITHAL ARC

THE circumzenithal arc is a *halo phenomenon* that appears like a multicoloured smile in the sky. Photographs of it look as if some fool's got a rainbow snap upside down, but this bow of colours actually appears in a totally different part of the sky from rainbows. On the 25 or so times a year that it appears,* it forms high up in the sky, like the fragment closest to the Sun of a circle around the zenith (directly up).

The ice crystals of Cirrus have found something to smile about.

Whenever you notice the spots of light on either side of the Sun called sundogs (p95), always look directly up** because you might also be able to add this most beautiful of all halo phenomena to your collection of cloud optical effects, for it is produced by the same cloud ice crystals. It appears as sunlight is *refracted* by the ice crystals of thin layers of high clouds, such as Cirrus (p19), Cirrostratus (p23) and Cirrocumulus (p21), or the ground-level ice-crystal cloud, diamond dust (p75).

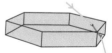

Circumzenithal arcs can appear when a cloud's ice crystals are in the shape of hexagonal plates and aligned more or less horizontally (like falling autumn leaves). As the light shines through the tops of the crystals and exits through a side face, the colours are separated and the combined sparkles appear as an arc.

* See note about frequency on p112.
** See p112 for halo-spotting safety advice.

The best sun pillars appear just after the Sun has dipped below the horizon.

| ☐ **I spotted sun pillars** | Points scored: 35 + 10 | Add to page 7 |

Date: .... Time: ....

Location: ....

Weather conditions: ....

☐ Photographed   Image file(s): ....

Cloud-collecting Points:
Any sun/moon pillars .... 35 points ☐
Bonus for a pillar below the Sun, seen
from a plane, which contains a bright
patch, called a subsun ..... 10 points ☐

Seen in: Cirrostratus (p23), Cirrus (p19),
Cirrocumulus (p21), diamond dust (p75).
Sun's elevation: visible when Sun is less
than 5° above horizon; clearest when
Sun is 2° below horizon.

SUN PILLARS

SUN pillars are vertical streaks of light that appear above and below a low Sun as it shines through ice-crystal clouds, such as Cirrus (p19), Cirrostratus (p23) and Cirrocumulus (p21), or the ground-level ice-crystal cloud, diamond dust (p75). At night, they are called 'moon pillars'.

Like many halo phenomena, pillars can also be seen by the light of a full Moon.

These *halo phenomena*, which appear on about 25 days of the year,* are due to sunlight reflecting off the surface of ice crystals. They are akin to 'glitter paths' that shine on the rippled surface of the sea. The pillar extending above the Sun appears brightest when the Sun is just below the horizon.

forms upper pillar

forms lower pillar

Most halo phenomena look best when the clouds' crystals are optically pure, regularly shaped and neatly aligned, but this is not the case for sun pillars. The light needs only to glance off a surface, so the crystals can be rough, irregular and jumbled. The poor man's halo phenomena, they often appear when cloud crystals aren't quite right for the more refined arcs, rings and spots of light to form (pp93–97).

crystals wobble or spin

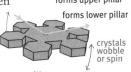

Sun pillars form when the sunlight reflects off the surface of the cloud's ice crystals. These need to be wobbling or rotating about a horizontal axis as they fall through the air for tall pillars to appear.

* See note about frequency on p112.
** See p112 for halo-spotting safety advice.

TECHNICAL TERMS

It's hard to explain clouds without sometimes using technical language. When the cloud entries have required this, the term has been written in italics and an explanation with respect to clouds is given below. Where explanations refer to other technical terms, these appear here in italics too.

Accessory clouds: Those that form close to one of the main cloud types, with which they sometimes merge. (See pp43–47.)

Cloudlets: An extended layer of cloud can be either smooth and continuous or made up of discernible clumps of cloud. These clumps, which can be joined or separate, are cloudlets.

Cloud levels: For the purpose of classifying clouds, meteorologists divide the *troposphere* into three arbitrary levels, or 'étages': low, medium and high. Clouds associated with each level are those whose bases are typically found within its altitude range.

Condensation nuclei: The plentiful tiny particles, floating in the atmosphere, that are the right size and shape to act as 'seeds' on which *water vapour* can start to condense to form cloud droplets. Just a few thousandths of a millimetre in size, they might be particles of salt (from the sea), dust (minerals, dead vegetation), ash or man-made pollution.

Convection clouds: Those that are formed by rising *convection currents* of air. By expanding as they rise, the currents can cool enough for some of their *water vapour* to condense into cloud droplets. (See pp9, 27).

Convection currents: Air rising (or sinking) as a result of being heated (or cooled) and therefore becoming less dense (or denser) than the surrounding air. An example is a thermal, when air, warmed by the Sun-baked ground, floats upwards.

Diffraction: How light waves spread out as they pass around tiny obstacles, such as cloud droplets or ice crystals. Different wavelengths are spread by different degrees and, if the cloud particles are of uniform size, each wavelength can produce an interference pattern of bright and dark fringes. The result is bands of different coloured light. (See pp81, 83, 87, 89, 91.)

Genera of cloud: The ten main classifications of cloud (see pp9–27). Clouds can belong to just one genus at a time.

Halo phenomena: Optical effects that result from the reflection and *refraction* of light as it passes through ice-crystal clouds. When optically pure and regularly shaped, the crystals can act as tiny prisms that bend or disperse the light. Those at certain angles from the Sun are more likely than others to shine light at an observer. Their combined sparkles are what produce white or coloured arcs, rings and spots of light across the sky. (See pp93–99.)

Icing nuclei: Particular airborne particles that can act as the 'seeds' on to which liquid cloud droplets can start to freeze into ice crystals. While there are generally plenty of *condensation nuclei* around, icing ones can be quite scarce. Without them, droplets can remain 'supercooled' until temperatures drop to around –40°C. A lack of suitable icing nuclei is why clouds often remain in the form of *supercooled droplets* below 0°C.

Mesosphere: The region of the atmosphere that is above the *stratosphere*, with an altitude range from 30 to over 50 miles. It's the coldest part of the atmosphere, where temperatures drop to –125°C, and home to the Earth's highest clouds (see p79).

Multicell storm: A thunderstorm, in which more than one 'cell' is active at once. Each cell is a system of rising and falling air currents, which combine to form a large Cumulonimbus (p27) cloud structure. As air rushes in at the base to feed the vertical growth of one cell, it can trigger another cell to start building at the front of the storm, which appears as a rising cloud tower. Though less co-ordinated than in a *supercell storm*, this succession of cells can extend the storm's duration for many hours.

Refraction: The way light changes direction as it passes between air and the water droplets or ice crystals of a cloud. Since different wavelengths are bent by different degrees, this generally leads to the sunlight being separated into its constituent colours. (See pp89, 93–97.)

Species of cloud: Cloud classification, by which some of the ten main cloud types, or *genera*, are subdivided. A cloud of a particular genus can belong to just one species at a time. For a table of all the formal cloud classifications, see p104.

Stable air: A region of the atmosphere in which a parcel of air tends to sink back down or rise back up to its original level when displaced, due to the way temperature varies with altitude.

Stratosphere: The region of the atmosphere above the *troposphere* and below the *mesosphere*, with an altitude range of 12–30 miles. It is separated from the turbulent air currents nearer the ground by a *temperature inversion*, called the *tropopause*. The only clouds that enter the stratosphere are nacreous (p77) and the tops of huge Cumulonimbus (p27).

Supercell storm: A very large, violent and persistent thunderstorm, consisting of a self-organising structure of Cumulonimbus cloud (p27). The storm's up- and down-draughts become co-ordinated to feed and maintain the cloud's structure, often for several hours and over long distances. Supercell storms can produce strong winds, frequent lightning, torrential rain, large hailstones and tornadoes.

Supercooled droplets: Cloud droplets that refuse to freeze, even though they are below 0°C. Mid-level clouds, such as

Altocumulus (p15), are often supercooled. Unless there are enough *icing nuclei* to facilitate freezing, water can remain in droplet form at temperatures as low as −40°C.

Supplementary features: Cloud forms that only appear attached to one of the main types of cloud, or *genera*. (See pp49–57.)

Temperature inversion: Normally, the air gets colder as you rise through the *troposphere*. An inversion is when, within a certain altitude range, this fall-off in temperature is arrested, or temperatures actually increase with height. Such a temperature profile acts as a 'thermal ceiling' that tends to halt the vertical growth of clouds. Inversions can occur at any altitude in the *troposphere*, the top of which is defined in terms of an inversion, called the *tropopause*. This is what causes large Cumulonimbus clouds to splay out in an anvil shape (see pp27, 57).

Tropopause: The *temperature inversion* that marks the boundary between the *troposphere* and the *stratosphere*.

Troposphere: The lower region of the atmosphere, up to about eight miles in the mid-latitudes, higher at the tropics and lower at the poles. This is the turbulent region where Earth's weather is generated. Almost all clouds form within the troposphere, the two exceptions being nacreous (p77) and noctilucent (p79).

Unstable air: A region of the atmosphere in which a parcel of air displaced up- or downwards has a tendency to keep going, due to the way the temperature varies with altitude.

Varieties of cloud: A visual characteristic that's used in the classification of clouds. Any cloud can exhibit several of these at once, and so belong to more than one variety at a time. For a table of all the official cloud classifications, see p104.

Water vapour: The gaseous state of water. Clouds appear when this invisible gas, one of the most variable components of air, forms into liquid droplets or solid ice crystals, which are visible.

CLOUD CLASSIFICATION

Clouds are officially classified according to a Latin Linnean system (like the one used for plants and animals), based on their heights and appearance. While the system isn't comprehensive, most clouds fall into one of the ten basic *genera*. They can further be defined as one of the possible

Genus	Species *(Can only be one)*	Varieties *(Can be more than one)*	Accessory Clouds & Supp. Features	LEVEL see p104
Cumulus *(9)*	humilis mediocris congestus fractus	radiatus *(39)*	pileus *(43)* velum *(45)* virga *(51)* praecipitatio arcus *(53)* pannus *(47)* tuba *(55)*	LOW CLOUDS
Stratocumulus *(11)*	stratiformis lenticularis *(29)* castellanus *(33)*	translucidus perlucidus opacus duplicatus *(41)* undulatus *(35)* radiatus *(39)* lacunosus *(37)*	mamma *(49)* virga *(51)* praecipitatio	LOW CLOUDS
Stratus *(13)*	nebulosus fractus	opacus translucidus undulatus *(35)*	praecipitatio	LOW CLOUDS
Altocumulus *(15)*	stratiformis lenticularis *(29)* castellanus *(33)* floccus	translucidus perlucidus opacus duplicatus *(41)* undulatus *(35)* radiatus *(39)* lacunosus *(37)*	virga *(51)* mamma *(49)*	MID-LEVEL

species, and any of the possible *varieties*. *Accessory clouds* and *supplementary features* are ones often found near particular genera. Genera are spelt with a capital letter. The numbers in brackets refer to the relevant pages.

	Genus	Species *(Can only be one)*	Varieties *(Can be more than one)*	Accessory Clouds & Supp. Features
			translucidus	virga *(51)*
			opacus	praecipitatio
MID LEVEL	Altostratus *(17)*	(none)	duplicatus *(41)*	pannus *(47)*
			undulatus *(35)*	mamma *(49)*
			radiatus *(39)*	
		fibratus *(31)*	intortus	
		uncinus	radiatus *(39)*	
	Cirrus *(19)*	spissatus	vertebratus	mamma *(49)*
		castellanus *(33)*	duplicatus *(41)*	
		floccus		
HIGH CLOUDS		stratiformis		
	Cirrocumulus *(21)*	lenticularis *(29)*	undulatus *(35)*	virga *(51)*
		castellanus *(33)*	lacunosus *(37)*	mamma *(49)*
		floccus		
	Cirrostratus *(23)*	fibratus *(31)*	duplicatus *(41)*	(none)
		nebulosus	undulatus *(35)*	
LOW CLOUDS	Nimbostratus *(25)* *(extends through more than one level)*	(none)	(none)	praecipitatio
				virga *(51)*
				pannus *(47)*
MULTI-LEVEL CLOUDS	Cumulonimbus *(27)* *(extends through all three levels)*	calvus capillatus	(none)	praecipitatio
				virga *(51)*
				pannus *(47)*
				incus *(57)*
				mamma *(49)*
				pileus *(43)*
				velum *(45)*
				arcus *(53)*
				tuba *(55)*

PHOTOGRAPHERS INDEX

We are very grateful to all the members of The Cloud Appreciation Society who agreed to contribute clouds from their collections to illustrate this handbook. All image copyrights remain with the photographers.

Amundsen, Håkon S. 76
Annesley, John 85
   www.annesleyphoto.com
Ansell, David 62
Atkinson, Richard 64
Battle, Angela 98
Bennett, Linda 89
Binnewies, Stefan 90
   www.capella-observatory.com
Bluemel, Andrew 30, 37
Blyth, Lissa C. 49
Brookes, Kate 36
Burgess, Anne 40
Burt, Stephen 20, 21, 33, 41, 46, 47
Chaudhry, Bilal 77
Childs, Gregg 15
Cho, Hong Gyu 81
Chudleigh, Brian 38
Clark, Matt 73
Clarke, Eunice 8
Clover, Jackie 51
Cooper, Paul 39
Costagliola, Matteo 37
Counihan, Nigel 95
Craggs, Angela 84
Dahl, Terje 77

Davenport, Michael 58
Dossler, Thomas 74
Drayton, Robin 97
Fairweather, Nathan 72
Farago, Zoltan 12
   www.flickr.com/
   photos/25637831@N07
Feci, Antonio 11
Gardner, Jo 70
Gerneke, Dane 80
Goddard, Ian 89
Goloy, Giselle 60
Greaves, Andrew 65
Gruber, Michèle 13
Harrison, Vicki 68
Harsch, Claudia 10
Herd-Smith, Bernardo 27
Hicks, Jeff 67
   www.j3ff.co.uk
Hightower, Bryan 5
Hotte, Bob 9
Howie, Frank 18
Hunter, Paul 73
Jack, Meggan 83
Justice, Jodie 67
Karanik, Jim 55

Keen, Graham 23
Kirk, Andrew 34
Kordaso, Tomislav 15
Kuitenbrouwer, Frits 61
Leech, Dave 25
Lorent, Guy 63
Lovibond, Nick 91
Loxley, Ian 16, 23, 24
Lučić, Dunja 29
Machin, Richard B. 29
MacPherson, John 85
   www.john.macpherson.btinternet.
   co.uk/johnmacphersonph.html
Malone, Sandra 42
Mann, Peter 35
Marks, Richard 19
McArthur, Gary 65
McVity, Stuart 94
Mokhtar, Ruziana Mohd 11
Moore, Shannon D. 31
   www.OutdoorPhoto.com
Moore, Tina 14
Murray, Joseph 71
Newton, Dave 86
Ohrberg, Mick 26
Oste, Jurgen 50
Pareto, Alberto 53
Paterson, John 96
Pellin, Luis Antonio Gil 44, 57
Penolazzi, Enrico 51, 81
Peterson, Bob 59
Pfeifer, Katrin 35
Pfister, Reto 54

Pleijel, Håkan 79
Pöpsel, Josef 90
   www.capella-observatory.com
Powell, Ginnie 56
Pretor-Pinney, Gavin 66
Priaulx, Laura 17
Rehwald, Eric 71
Robinson, Terry 61
Rowlands, John 49
Russo, Alan 53, 89
Sale, Norwynne 48
Santikarn, Sitthivet 88, 93
Schrayer, Grover 75
   www.flickr.com/
   photos/14833125@N02
Seccombe, Wally 19
Seckold, Robert 87
Smith, Janice 9
Summers, Rachel 69
Summey, Ally 82
Tack, Jason 22
Valentine, Bill 78
Verwest, Ryan 28, 59
Warren, Celia 32
Westmaas, Ron 27
Wheelhouse, Sue 43
Williams, Elizabeth 92
Williams, Geoff 45
Wolck, Nancy 52
Zenko, Peg 83, 99
   www.tangentphotos.com
Zinkova, Mila 87

CLOUD IMAGE INDEX

When parts of the sky bear similarities to these thumbnails, refer to the page numbers to see what the cloud types might be.

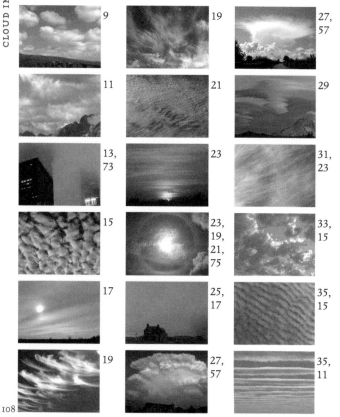

9

19

27, 57

11

21

29

13, 73

23

31, 23

15

23, 19, 21, 75

33, 15

17

25, 17

35, 15

19

27, 57

35, 11

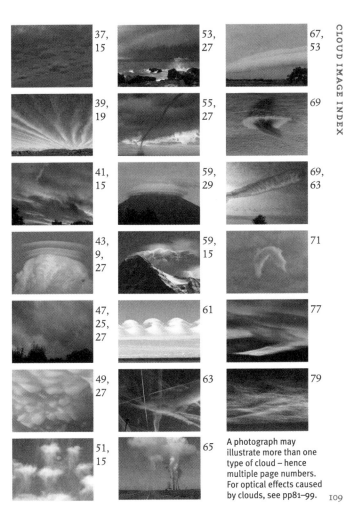

37, 15

39, 19

41, 15

43, 9, 27

47, 25, 27

49, 27

51, 15

53, 27

55, 27

59, 29

59, 15

61

63

65

67, 53

69

69, 63

71

77

79

A photograph may illustrate more than one type of cloud – hence multiple page numbers. For optical effects caused by clouds, see pp81–99.

The main entries for clouds are in bold.

accessory clouds 27, 43, 45, 100, 104-105

Altocumulus 8, 10-11, **14-15**, 17-18, 20-21, 28-29, 32-36, 38, 40-41, 48, 50, 61, 80-82, 86, 90, 103-104

Altostratus 12-13, **16-17**, 22-25, 34, 38, 40-41, 44, 46, 48, 50, 80-82, 86, 90, 105

anvil, clouds in the shape of 19, 27, 48-49, 57, 102

arcus 27, **52-53**, 66-67, 104-105

banner cloud **58-59**

Brocken spectre 86-87

cap cloud 42-43, **58-59**

castellanus 11, 14-15, 19, 21, **32-33**, 104-105

circumzenithal arc 88, **96-97**

Cirrocumulus 14-15, **20-21**, 28, 32-34, 36-37, 48, 50, 63, 80-82, 92-97, 99, 105

Cirrostratus 12, 16-17, 20-21, **22-23**, 30-31, 34, 40-41, 62-63, 75, 78, 82-83, 92-99, 105

Cirrus 18, **19-20**, 27, 29-33, 38-41, 48-50, 60-63, 68-69, 75, 92-99, 105

cloudbow **90-91**

cloudlet 11, 15, 20-21, 35, 37, 100

cloud levels 29, 31, 33, 37, 39, 61, 69, 100

cloud streets 9, 39

condensation nuclei 65, 100-101

contrail 38, 48, **62-63**, 65, 76, 94

convection clouds 43-45, 89, 100

convection currents 65, 100-101

corona **81-83**, 86, 92

crepuscular rays 72, **84-85**

Cumulonimbus 8-9, 19, 24-25, **26-27**, 33, 43-46, 48-49, 52-57, 88, 102-103, 105

Cumulus **8-9**, 10, 15, 21, 26-28, 33, 38-39, 42-46, 50, 52-55, 57, 84-85, 88, 102, 104

diamond dust **74-75**, 92-99

diffraction, of sunlight by cloud droplets and crystals 76-77, 81, 83, 87, 91, 101

distrail 62-3, 69

duplicatus 11, 15, 17, 19, 23, 34, **40-41**, 104-105

fair-weather clouds 8-9

fallstreak hole 51, **68-69**

fallstreaks 19

fibratus 19, 22-23, **30-31**, 40-41, 105

fog 13, 60-61, 64, **72-73**, 75, 86-87, 90-91

fogbow 90-91

fumulus 65

funnel cloud 55

glory 66-67, 82-83, **86-87**, 90-91

gust front 53

halo, 22° 83, **92-93**, 95

halo phenomena 17, 19, 22-23, 74-75, 88-89, 93, 95, 97, 99, 101, 112

holes in clouds 11, 15, 21, 36-37, 104-105

  see also fallstreak hole

hole-punch clouds 69

horseshoe vortex **70-71**

icing nuclei 63, 69, 101, 103

incus 27, 49, 53, **56-57**, 105

inversion, temperature 57, 101-103
iridescence 68, 76-77, **80-81**, 83
irisation 81
glitter paths 99
jellyfish (clouds that look like) 15, 51, 68
jet-stream Cirrus 38-39
Kelvin-Helmholtz **60-61**
King of Clouds 26
lacunosus 11, 15, 21, **36-37**, 104-105
landspouts 55, 65
lenticularis 11, 15, 21, **28-29**, 42-43, 59-60, 77, 80-81, 104-105
mamma 27, **48-49**, 104-105
mesosphere 79, 102-103
mist *see* fog
mother-of-pearl clouds 77, 81
Morning Glory cloud 66-67
nacreous **76-77**, 80-81, 103
Nimbostratus 12, 17, 19, **24-25**, 27, 46, 50, 105
noctilucent 34, 36, **78-79**, 103
pannus 13, 25, **46-47**, 104-105
perlucidus 11, 15, 104
perspective effect 39, 85, 87
pile d'assiettes 28-29
pileus 27-28, 37, **42-43**, 45, 58, 80-81, 104-105
praecipitatio 51, 104-105
pyrocumulus 48, **64-65**
radiatus 9, 11, 15, 17, 19, 30, **38-39**, 104-105
rainbow 23, 87, **88-89**, 90-91, 97
reflection, of sunlight in cloud droplets and crystals 19, 23, 75, 87, 89, 91, 99, 101
refraction, of sunlight by cloud droplets and crystals 19, 23, 75, 87, 89, 91, 93, 95, 97, 101-102
roll cloud 53, **66-67**

shelf cloud 52-53
species of cloud 8-9, 11, 13, 15, 17-19, 21, 23, 25, 27-34, 36, 38, 40-41, 59, 77, 102, 104-105
stable air 29, 43, 45, 59, 77, 102
storm:
    multicell 27, 52-55, 102
    supercell 27, 52-55, 71, 103
Stratocumulus 8, **10-11**, 14, 28-29, 32-41, 48, 50, 58-59, 61, 80, 82, 84-85, 90, 104
stratosphere 57, 77, 81, 102-103
Stratus 10, **12-13**, 16-17, 22, 24-25, 34, 47, 58-59, 73, 82, 86, 90, 104
subsun 98
sundogs **94-95**, 96-97
sun pillars **98-99**
supercooled droplets 51, 63, 69, 101, 103
supplementary features 27, 49, 51, 53, 57, 103, 105
tropopause 57, 102-103
troposphere 19, 31, 39, 57, 76, 100-103
tuba 27, 53, **54-55**, 65, 71, 104-105
UFO-shaped clouds 29
undulatus 11, 13, 15, 17, 21, 23, **34-35**, 38, 60-61, 104-105
unstable air 9, 33, 103
varieties of cloud 13, 35, 37, 39-41, 61, 103
velum 27, 43, **44-45**, 104-105
virga 15, 18, 21, **50-51**, 68-69, 104-105
vortices 54-55, 61, 70-71
waterspouts 54, 55, 65
water vapour 63, 73, 75, 100, 103
waves, in clouds 11, 13, 15, 17, 21, 23, 29, 35, 53, 60-61, 67, 75, 77, 81, 101-102
wind shear, cloud forms caused by 35, 61, 71

You can become a member of The Cloud Appreciation Society by visiting www.cloudappreciationsociety.org. The website has a gallery where you can post your prized cloud images and advice about photographing clouds.

Anyone who wants to know more about cloud optical phenomena should visit the Atmospheric Optics website run by Les Cowley (member 14) at www.atoptics.co.uk.

If you're interested in finding out about the weather in general, you might consider joining the Royal Meteorological Society. It publishes a monthly journal, *Weather*, and holds meetings every year for those interested in weather and climate. You can find out more at www.rmets.org.

The cloud charts on the endpapers were illustrated by Anthony Haythornthwaite (member 2367) who can be contacted at anthony@aqhthestudio.co.uk.

A cautionary note about observing halo phenomena:

Ice-crystal clouds that give rise to halo phenomena (see pp93–99) block very little of the Sun's rays. Always shield both eyes from the Sun when looking at them. It's a good idea to stand so that a building, a tree or a nearby giraffe shields the Sun itself from your vision. Especial care should be taken when photographing halo phenomena with an SLR camera, as the viewfinder will tend to increase the chance of retinal damage from direct rays.

Frequency of optical phenomena:

All figures of the frequency of occurrence of haloes and other optical phenomena are based on observations made over Germany, between 1986 and 2004, that were compiled by the German 'Meteor Work Group': www.meteoros.de.

First published in Great Britain in 2009 by Sceptre
An imprint of Hodder & Stoughton, An Hachette UK company.

1

Copyright © Gavin Pretor-Pinney 2009
Copyright of the individual photographs remains with the credited photographers.

The right of Gavin Pretor-Pinney to be identified as the Author of the Work has been asserted by him in accordance with the Copyright, Designs and Patents Act 1988.

All rights reserved. No part of this publication may be reproduced, stored in a retrieval system, or transmitted, in any form or by any means without the prior written permission of the publisher, nor be otherwise circulated in any form of binding or cover other than that in which it is published and without a similar condition being imposed on the subsequent purchaser.

A CIP catalogue record for this title is available from the British Library
ISBN 978 0 340 91943 9

Designed and typeset in Dante, Meta and Base Twelve by Gavin Pretor-Pinney.

Hodder & Stoughton policy is to use papers that are natural, renewable and recyclable products and made from wood grown in sustainable forests. The logging and manufacturing processes are expected to conform to the environmental regulations of the country of origin.

Hodder & Stoughton Ltd
338 Euston Road
London NW1 3BH

 Printed in Italy by L.E.G.O. Spa.

www.hodder.co.uk

45,000 f
CIRRUS
40,000 f
35,000 f
30,000 f
CIRROSTRATUS
25,000 f
20,000 f
ALTOSTRATUS
15,000 f
10,000 f
NIMBOSTRATUS
STRATOCUMULUS
5,000 f
STRATUS
0 f